Displaying Time Series, Spatial, and Space-Time Data with R

Second Edition

T0138625

Chapman & Hall/CRC The R Series

Series Editors

John Chambers, Stanford University
Torsten Hothorn, University of Zurich, Switzerland
Duncan Temple Lang, University of California, Davis
Hadley Wickham, RStudio

Basics of Matrix Algebra for Statistics with R
Nick Fieller

Introductory Fisheries Analyses with R
Derek H. Ogle

Statistics in Toxicology Using R
Ludwig A. Hothorn

Spatial Microsimulation with R
Robin Lovelace, Morgane Dumont

Extending R
John M. Chambers

Using the R Commander
A Point-and-Click Interface for R
John Fox

Computational Actuarial Science with R
Arthur Charpentier

bookdown
Authoring Books and Technical Documents with R Markdown
Yihui Xie

Testing R Code
Richard Cotton

R Primer
Second Edition
Claus Thorn Ekstrøm

Flexible Regression and Smoothing
Using GAMLSS in R
Mikis D. Stasinopoulos, Robert A. Rigby, Gillian Z. Heller, Vlasios Voudouris, and Fernanda De Bastiani

The Essentials of Data Science
Knowledge Discovery Using R
Graham J. Williams

Blogdown
Creating Websites with R Markdown
Yihui Xie, Alison Presmanes Hill, Amber Thomas

Handbook of Educational Measurement and Psychometrics Using R
Christopher D. Desjardins, Okan Bulut

Displaying Time Series, Spatial, and Space-Time Data with R
Second Edition
Oscar Perpiñán Lamigueiro

For more information about this series, please visit: https://www.crcpress.com/go/the-r-series

Displaying Time Series, Spatial, and Space-Time Data with R

Second Edition

Oscar Perpiñán Lamigueiro

CRC Press
Taylor & Francis Group
Boca Raton London New York

CRC Press is an imprint of the
Taylor & Francis Group, an **informa** business

A CHAPMAN & HALL BOOK

CRC Press
Taylor & Francis Group
6000 Broken Sound Parkway NW, Suite 300
Boca Raton, FL 33487-2742

Printed on acid-free paper
Version Date: 20180629

International Standard Book Number-13: 978-1-1380-8998-3 (Hardback)

Library of Congress Cataloging-in-Publication Data

Names: Perpiñán Lamigueiro, Oscar, author.
Title: Displaying time series, spatial, and space-time data with R / Oscar Perpiñán Lamigueiro.
Description: Second edition. | Boca Raton : CRC Press, Taylor & Francis Group, 2018. | Includes bibliographical references and index.
Identifiers: LCCN 2018016574 | ISBN 9781138089983 (hardback)
Subjects: LCSH: Time-series analysis--Data processing. |
R (Computer program language)
Classification: LCC QA280 .P475 2018 | DDC 519.5/502855133--dc23
LC record available at https://lccn.loc.gov/2018016574

**Visit the Taylor & Francis Web site at
http://www.taylorandfrancis.com**

**and the CRC Press Web site at
http://www.crcpress.com**

Contents

CONTENTS

Chapter 1

Introduction

1.1 What This Book Is About

A data graphic is not only a static image but also tells a story about the data. It activates cognitive processes that are able to detect patterns and discover information not readily available with the raw data. This is particularly true for time series, spatial, and space-time datasets.

There are several excellent books about data graphics and visual perception theory, with guidelines and advice for displaying information, including visual examples. Let's mention *The Elements of Graphical Data* (Cleveland 1994) and *Visualizing Data* (Cleveland 1993) by W. S. Cleveland, *Envisioning Information* (Tufte 1990) and *The Visual Display of Quantitative Information* (Tufte 2001) by E. Tufte, *The Functional Art* by A. Cairo (Cairo 2012), and *Visual Thinking for Design* by C. Ware (Ware 2008). Ordinarily, they do not include the code or software tools to produce those graphics.

On the other hand, there is a collection of books that provides code and detailed information about the graphical tools available with R. Commonly they do not use real data in the examples and do not provide advice for improving graphics according to visualization theory. Three books are the unquestioned representatives of this group: *R Graphics* by P. Murrell (Murrell 2011), *Lattice: Multivariate Data Visualization with R* by D. Sarkar (Sarkar 2008), and *ggplot2: Elegant Graphics for Data Analysis* by H. Wickham (Wickham 2016).

1

This book proposes methods to display time series, spatial, and space-time data using R, and aims to be a synthesis of both groups providing code and detailed information to produce high-quality graphics with practical examples.

1.2 What You Will *Not* Find in This Book

- **This is not a book to learn R.**

  Readers should have a fair knowledge of programming with R to understand the book. In addition, previous experience with the zoo, sp, raster, lattice, ggplot2, and grid packages is helpful.

  If you need to improve your R skills, consider these information sources:

  - Introduction to R[1].
  - Official manuals[2].
  - Contributed documents[3].
  - Mailing lists[4].
  - R-bloggers[5].
  - Books related to R[6] and particularly *Software for Data Analysis* by John M. Chambers (Chambers 2008).

- **This book does not provide an exhaustive collection of visualization methods.**

  Instead, it illustrates what I found to be the most useful and effective methods. Notwithstanding, each part includes a section titled "Further Reading" with bibliographic proposals for additional information.

- **This book does not include a complete review or discussion of R packages.**

---

[1]http://cran.r-project.org/doc/manuals/R-intro.html
[2]http://cran.r-project.org/manuals.html
[3]http://cran.r-project.org/other-docs.html
[4]http://www.r-project.org/mail.html
[5]http://www.r-bloggers.com
[6]http://www.r-project.org/doc/bib/R-books.html

Their most useful functions, classes, and methods regarding data and graphics are outlined in the introductory chapter of each part, and conveniently illustrated with the help of examples. However, if you need detailed information about a certain aspect of a package, you should read the correspondent package manual or vignette. Moreover, if you want to know additional alternatives, you can navigate through the CRAN Task Views about Time Series[7], Spatial Data[8], Spatiotemporal Data[9], and Graphics[10].

- **Finally, this book is not a handbook of data analysis, geostatistics, point pattern analysis, or time series theory**.

  Instead, this book is focused on the exploration of data with visual methods, so it may be framed in the Exploratory Data Analysis approach. Therefore, this book may be a useful complement for superb bibliographic references where you will find plenty of information about those subjects. For example, (Chatfield 2016), (Cressie and C. Wikle 2015), (Slocum 2005) and (Bivand, E. J. Pebesma, and Gomez-Rubio 2013).

1.3 How to Read This Book

This book is organized into three parts, each devoted to different types of data. Each part comprises several chapters according to the various visualization methods or data characteristics. The chapters are structured as independent units so readers can jump directly to a certain chapter according to their needs. Of course, there are several dependencies and redundancies between the sets of chapters that have been conveniently signaled with cross-references.

The content of each chapter illustrates how to display a dataset starting with an easy and direct approach. Often this first result is not entirely satisfactory so additional improvements are progressively added. Each step involves additional complexity which, in some cases, can be overwhelming during a first reading. Thus, some sections, marked with the sign ✿, can be safely skipped for later reading.

Although I have done my best to help readers understand the methods and code, you should not expect to understand it after one reading. The

---

[7]http://cran.r-project.org/web/views/TimeSeries.html

[8]http://cran.r-project.org/web/views/Spatial.html

[9]http://cran.r-project.org/web/views/SpatioTemporal.html

[10]http://cran.r-project.org/web/views/Graphics.html

key is practical experience, and the best way is to try out the code with the provided data **and** modify it to suit your needs with your own data. There is a website and a code repository to help you in this task.

1.3.1 Website and Code Repository

The book website with the main graphics of this book is located at

`http://oscarperpinan.github.com/bookvis`

The full code is freely available from the repository:

`https://github.com/oscarperpinan/bookvis`

On the other hand, the datasets used in the examples are either available at the repository or can be freely obtained from other websites. It must be underlined that the combination of code and data freely available allows this book to be fully reproducible.

I have chosen the datasets according to two main criteria:

- They are freely available without restrictions for public use.

- They cover different scientific and professional fields (meteorology and climate research, economy and social sciences, energy and engineering, environmental research, epidemiology, etc.).

The repository and the website can be downloaded as a compressed file[11], or if you use `git`, you can clone the repository with:

```
git clone https://github.com/oscarperpinan/bookvis.git
```

1.4   R Graphics

There are two distinct graphics systems built into R, referred to as traditional and grid graphics. Grid graphics are produced with the `grid` package (Murrell 2011), a flexible low-level graphics toolbox. Compared with the traditional graphics model, it provides more flexibility to modify or add content to an existent graphical output, better support for combining different outputs easily, and more possibilities for interaction. All the graphics in this book have been produced with the grid graphics model.

Other packages are constructed over it to provide high-level functions, most notably the `lattice` and `ggplot2` packages.

---

[11]`https://github.com/oscarperpinan/bookvis/archive/master.zip`

1.4.1   lattice

The `lattice` package (Sarkar 2008) is an independent implementation of Trellis graphics, which were mostly influenced by *The Elements of Graphing Data* (Cleveland 1994). Trellis graphics often consist of a rectangular array of panels. The `lattice` package uses a *formula* interface to define the structure of the array of panels with the specification of the variables involved in the plot. The result of a `lattice` high-level function is a `trellis` object.

For bivariate graphics, the formula is generally of the form y ~ x representing a single panel plot with y versus x. This formula can also involve expressions. The main function for bivariate graphics is `xyplot`.

Optionally, the formula may be y ~ x | g1 * g2 and y is represented against x conditional on the variables g1 and g2. Each unique combination of the levels of these conditioning variables determines a subset of the variables x and y. Each subset provides the data for a single panel in the Trellis display, an array of panels laid out in columns, rows, and pages.

For example, in the following code, the variable `wt` of the dataset `mtcars` is represented against the `mpg`, with a panel for each level of the categorical variable `am`. The points are grouped by the values of the `cyl` variable.

```
xyplot(wt ~ mpg | am, data = mtcars, groups = cyl)
```

For trivariate graphics, the formula is of the form z ~ x * y, where z is a numeric response, and x and y are numeric values evaluated on a rectangular grid. Once again, the formula may include conditioning variables, for example z ~ x * y | g1 * g2. The main function for these graphics is `levelplot`.

The plotting of each panel is performed by the panel function, specified in a high-level function call as the `panel` argument. Each high-level `lattice` function has a default panel function, although the user can create new Trellis displays with custom panel functions.

`lattice` is a member of the recommended packages list so it is commonly distributed with R itself. There are more than 250 packages depending on it, and the most important packages for our purposes (`zoo`, `sp`, and `raster`) define methods to display their classes using `lattice`.

On the other hand, the `latticeExtra` package (Sarkar and Andrews 2016) provides additional flexibility for the somewhat rigid structure of the Trellis framework implemented in `lattice`. This package complements the `lattice` with the implementation of layers via the `layer` function, and superposition of `trellis` objects and layers with the `+.trellis` function.

Using both packages, you can define a graphic with the formula interface (under the lattice model) and overlay additional content as layers (following the ggplot2 model).

1.4.2 ggplot2

The ggplot2 package (Wickham 2016) is an implementation of the system proposed in *The Grammar of Graphics* (Wilkinson 2005), a general scheme for data visualization that breaks up graphs into semantic components such as scales and layers. Under this framework, the definition of the graphic with ggplot2 is done with a combination of several functions that provides the components, instead of the formula interface of lattice.

With ggplot2, a graphic is composed of:

- A dataset, data, and a set of mappings from variables to aesthetics, aes.

- One or more layers, each composed of: a geometric object, geom_*, to control the type of plot you create (points, lines, etc.); a statistical transformation, stat_*; and a position adjustment (and optionally, additional dataset and aesthetic mappings).

- A scale, scale_*, to control the mapping from data to aesthetic attributes. Scales are common across layers to ensure a consistent mapping from data to aesthetics.

- A coordinate system, coords_*.

- Optionally, a faceting specification, facet_*, the equivalent of Trellis graphics with panels.

The function ggplot is typically used to construct a plot incrementally, using the + operator to add layers to the existing ggplot object. For instance, the following code (equivalent to the previous lattice example) uses mtcars as the dataset, and maps the mpg variable on the x-axis and the wt variable on the y-axis. The geometric object is the point using the cyl variable to control the color. Finally, the levels of the am variable define the panels of the graphic.

```
ggplot(mtcars, aes(mpg, wt)) +
    geom_point(aes(colour=factor(cyl))) +
    facet_grid(. ~ am)
```

This package is very popular, with a large list packages depending on it. In the context of this book, time series can be displayed with it because the zoo package defines the autoplot function based on ggplot2. Regarding spatial data, recent versions of this package provide a geom function designed for spatial data. Detailed information is provided in Section 7.1.2.

1.4.3 Comparison between lattice and ggplot2

Which package to choose is, for a wide range of datasets, a question of personal preferences. You may be interested in a comparison between them published in a series of blog posts[12]. Consequently, where possible most of the code contains alternatives defined both with lattice and with ggplot2.

It is important to note that both latticeExtra and ggplot2 defined a function named layer. The ggplot2::layer function is rarely called by the user, because the wrapper functions geom_* and stats_ are preferred. On the other hand, the latticeExtra::layer function is designed to be directly called by the user, and therefore its masking must be prevented. Consequently, when the latticeExtra and ggplot2 packages are to be working together in the same session, the latticeExtra package must be loaded after ggplot2.

1.4.4 Interactive graphics

Both lattice and ggplot2 (and every package based on grid) generate static graphics. However, interactive web graphics produced with R have experienced a boost in recent years, mainly thanks to the package html-widgets (Vaidyanathan et al. 2017). This package provides a framework for creating R bindings to JavaScript libraries. This package is the base for important visualization packages such as dygraphs, highcharter, plotly, leaflet and mapview. They will be covered along the chapters of the book.

On the other hand, the package gridSVG (Murrell and Potter 2017) converts any grid scene to a Scalable Vector Graphics (SVG) document. The grid.hyperlink function allows a hyperlink to be associated with any component of the scene, the grid.animate function can be used to animate any component of a scene, and the grid.garnish function can be used to add SVG attributes to the components of a scene. By setting event

---

[12]http://learnr.wordpress.com/2009/06/28/

7

handler attributes on a component, plus possibly using the grid.script function to add JavaScript to the scene, it is possible to make the component respond to user input such as mouse clicks.

1.5 Packages

Throughout the book, several R packages are used. All of them are available from CRAN, and you must install them before using the code. Most of them are loaded at the start of the code of each chapter, although some of them are loaded later if they are used only inside optional sections (marked with ✿). You should install the last version available at CRAN to ensure correct functioning of the code.

Although the introductory chapter of each part includes a section with an outline of the most relevant packages, some of them deserve to be highlighted here:

- zoo (Zeileis and Grothendieck 2005) provides infrastructure for time series using arbitrary classes for the time stamps (Section 2.1.1).

- sp (E. Pebesma 2012) and sf (E. Pebesma 2018) provide a coherent set of classes and methods for the major spatial data types: points, lines, polygons, and grids (Sections 7.1.1 and 7.1.2). spacetime (E. Pebesma 2012) defines classes and methods for spatiotemporal data, and methods for plotting data as map sequences or multiple time series (Section 14.1.1).

- raster (R. J. Hijmans 2017) is a major extension of gridded spatial data classes. It provides a unified access method to different raster formats, permitting large objects to be analyzed with the definition of basic and high-level processing functions (Sections 7.1.3 and 14.1.2). rasterVis (Perpiñán and R. Hijmans 2017) provides enhanced visualization of raster data with methods for spatiotemporal rasters (Sections 7.1.4 and 14.1.3).

1.6 Software Used to Write This Book

This book has been written using different computers running Debian GNU Linux and using several gems of open-source software:

- org-mode (Schulte et al. 2012), LATEX, and AUCTEX, for authoring text and code.

- R (R Development Core Team 2017) with Emacs Speaks Statistics (Rossini et al. 2004).

- GNU Emacs as development environment.

1.7 About the Author

During the past 18 years, my main area of expertise has been photovoltaic solar energy systems, with a special interest in solar radiation. Initially I worked as an engineer for a private company, and I was involved in several commercial and research projects. The project teams were partly integrated by people with low technical skills who relied on the input from engineers to complete their work. I learned how a good visualization output eased the communication process.

Now I work as a professor and researcher at the university. Data visualization is one of the most important tools I have available. It helps me embrace and share the steps, methods, and results of my research. With students, it is an inestimable partner in helping them understand complex concepts.

I have been using R to simulate the performance of photovoltaic energy systems and to analyze solar radiation data, both as time series and spatial data. As a result, I have developed packages that include several graphical methods to deal with multivariate time series (namely, solaR (Perpiñán 2012), meteoForecast (Perpiñán and Almeida 2015), and PVF (Pinho-Almeida, Perpiñán, and Narvarte 2015)) and space-time data (rasterVis (Perpiñán and R. Hijmans 2017)).

1.8 Acknowledgments

Writing a book is often described as a solitary activity. It is certainly difficult to write when you are with friends or spending time with your family,... although with three little children at home I have learned to write prose and code while my baby wants to learn typing and my daughters need help to share a family of dinosaurs.

Seriously speaking, solitude is the best partner of a writer. But when I am writing or coding I feel I am immersed in a huge collaborative network of past and present contributors. Piotr Kropotkin described it with the following words (Kropotkin 1906):

> Thousands of writers, of poets, of scholars, have laboured to increase knowledge, to dissipate error, and to create that atmosphere of scientific thought, without which the marvels of our century could never have appeared. And these thousands of philosophers, of poets, of scholars, of inventors, have themselves been supported by the labour of past centuries. They have been upheld and nourished through life, both physically and mentally, by legions of workers and craftsmen of all sorts.

And Lewis Mumford claimed (Mumford 1934):

> Socialize Creation! What we need is the realization that the creative life, in all its manifestations, is necessarily a social product.

I want to express my deepest gratitude and respect to all those women and men who have contributed and contribute to strengthening the communities of free software, open data, and open science. My special thanks go to the people of the R community: users, members of the R Core Development Team, and package developers.

With regard to this book in particular, I would like to thank John Kimmel for his constant support, guidance, and patience.

Last, and most importantly, thanks to Candela, Marina, and Javi, my crazy little shorties, my permanent source of happiness, imagination, and love. Thanks to María, *mi amor, mi cómplice y todo*.

Part I

Time Series

Chapter 2

Displaying Time Series: Introduction

A time series is a sequence of observations registered at consecutive time instants. When these time instants are evenly spaced, the distance between them is called the sampling interval. The visualization of time series is intended to reveal changes of one or more quantitative variables through time, and to display the relationships between the variables and their evolution through time.

The standard time series graph displays the time along the horizontal axis. Several variants of this approach can be found in Chapter 3. On the other hand, time can be conceived as a grouping or conditioning variable (Chapter 4). This solution allows several variables to be displayed together with a scatterplot, using different panels for subsets of the data (time as a conditioning variable) or using different attributes for groups of the data (time as a grouping variable). Moreover, time can be used as a complementary variable that adds information to a graph where several variables are confronted (Chapter 5).

These chapters provide a variety of examples to illustrate a set of useful techniques. These examples make use of several datasets (available at the book website) described in Chapter 6.

2.1 Packages

The CRAN Tasks View "Time Series Analysis" [1] summarizes the packages for reading, vizualizing, and analyzing time series. This section provides a brief introduction to the zoo and xts packages. Most of the information has been extracted from their vignettes, webpages, and help pages. You should read them for detailed information.

Both packages extensively use the time classes defined in R. The interested reader will find an overview of the different time classes in R in (Ripley and Hornik 2001) and (Grothendieck and Petzoldt 2004).

2.1.1 zoo

The zoo package (Zeileis and Grothendieck 2005) provides an S3 class with methods for indexed totally ordered observations. Its key design goals are independence of a particular index class and consistency with base R and the ts class for regular time series.

Objects of class zoo are created by the function zoo from a numeric vector, matrix, or a factor that is totally ordered by some index vector. This index is usually a measure of time but every other numeric, character, or even more abstract vector that provides a total ordering of the observations is also suitable. It must be noted that this package defines two new index classes, yearmon and yearqtr, for representing monthly and quarterly data, respectively.

The package defines several methods associated with standard generic functions such as print, summary, str, head, tail, and [(subsetting). In addition, standard mathematical operations can be performed with zoo objects, although only for the intersection of the indexes of the objects.

On the other hand, the data stored in zoo objects can be extracted with coredata, which drops the index information, and can be replaced by coredata<-. The index can be extracted with index or time, and can be modified by index<-. Finally, the window and window<- methods extract or replace time windows of zoo objects.

Two zoo objects can be merged by common indexes with merge and cbind. The merge method combines the columns of several objects along the union or the intersection of the indexes. The rbind method combines the indexes (rows) of the objects.

---

[1] http://CRAN.R-project.org/view=TimeSeries

The `aggregate` method splits a `zoo` object into subsets along a coarser index grid, computes a function (`sum` is the default) for each subset, and returns the aggregated `zoo` object.

This package provides four methods for dealing with missing observations:

1. `na.omit` removes incomplete observations.

2. `na.contiguous` extracts the longest consecutive stretch of non-missing values.

3. `na.approx` replaces missing values by linear interpolation.

4. `na.locf` replaces missing observations by the most recent non-NA prior to it.

The package defines interfaces to `read.table` and `write.table` for reading, `read.zoo`, and writing, `write.zoo`, zoo series from or to text files. The `read.zoo` function expects either a text file or connection as input or a `data.frame`. `write.zoo` first coerces its argument to a `data.frame`, adds a column with the index, and then calls `write.table`.

2.1.2   xts

The `xts` package (Ryan and Ulrich 2013) extends the zoo class definition to provide a general time-series object. The index of an `xts` object must be of a time or date class: `Date`, `POSIXct`, `chron`, `yearmon`, `yearqtr`, or `timeDate`. With this restriction, the subset operator `[` is able to extract data using the ISO:8601 [2] time format notation `CCYY-MM-DD HH:MM:SS`. It is also possible to extract a range of times with a `from`/`to` notation, where both from and to are optional. If either side is missing, it is interpreted as a request to retrieve data from the beginning, or through the end of the data object.

Furthermore, this package provides several time-based tools:

- `endpoints` identifies the endpoints with respect to time.

- `to.period` changes the periodicity to a coarser time index.

- The functions `period.*` and `apply.*` evaluate a function over a set of non-overlapping time periods.

---

[2]http://en.wikipedia.org/wiki/ISO_8601

2.2 Further Reading

- (Wills 2011) provides a systematic analysis of the visualization of time series, and a section of (Heer, Bostock, and Ogievetsky 2010) summarizes the main techniques to display time series.

- (Cleveland 1994) includes a section about time series visualization with a detailed discussion of the banking to 45° technique and the cut-and-stack method. (Heer and Agrawala 2006) propose the multiscale banking, a technique to identify trends at various frequency scales.

- (Few 2008; Heer, Kong, and Agrawala 2009) explain in detail the foundations of the horizon graph (Section 3).

- The *small multiples* concept (Sections 3.2 and 4.1) is illustrated in (Tufte 2001; Tufte 1990).

- Stacked graphs are analyzed in (Byron and Wattenberg 2008), and the ThemeRiver technique is explained in (Havre et al. 2002).

- (Cleveland 1994; Friendly and Denis 2005) study the scatterplot matrices (Section 4.1), and (D. B. Carr et al. 1987) provide information about hexagonal binning.

- (Harrower and Fabrikant 2008) discuss the use of animation for the visualization of data. (Few 2007) exposes a software tool resembling the Trendalyzer.

- The D3 gallery [3] shows several great examples of time-series visualizations using the JavaScript library D3.js.

---

[3] https://github.com/mbostock/d3/wiki/Gallery

Chapter 3

Time on the Horizontal Axis

The most frequent visualization method of a time series uses the horizontal axis to depict the time index. This chapter illustrates several variants of this approach to display multivariate time series: multiple time series with different scales, variables with the same scale, and stacked graphs. The last section provides examples and code for producing interactive visualizations.

Along this chapter these subjects are covered: small multiples, panel functions, aspect ratio, diverging palettes, horizon and stacked graphs, and interactive visualization.

The most relevant packages in this chapter are: zoo and xts for reading and arranging data as time series; colorspace for defining color palettes; and dygraphs, highcharter, plotly, and stream-graph for interactive visualization.

3.1 Time Graph of Variables with Different Scales

There is a variety of scientific research interested in the relationship among several meteorological variables. A suitable approach is to display the time evolution of all of them using a panel for each of the variables. The superposition of variables with different characteristics is not very useful (unless their values were previously rescaled), so this option is postponed for Section 3.2.

For this example we will use the eight years of daily data from the SIAR meteorological station located at Aranjuez (Madrid). This multivariate time series can be displayed with the xyplot method of lattice for zoo objects with a panel for each variable (Figure 3.1).

```
library(zoo)
load('data/aranjuez.RData')

## The layout argument arranges panels in rows
xyplot(aranjuez, layout = c(1, ncol(aranjuez)))
```

The package ggplot2 provides the generic method autoplot to automate the display of certain classes with a simple command. The package zoo provides an autoplot method for the zoo class with a result similar to that obtained with xyplot (Figure 3.2)

```
autoplot(aranjuez) + facet_free()
```

3.1.1 ✍Annotations to Enhance the Time Graph

These first attempts can be improved with a custom panel function that generates the content of each panel using the information processed by xyplot, or overlaying additional layers with autoplot. One of the main enhancements is to highlight certain time regions that fulfill certain conditions. The package latticeExtra provides a nice solution for xyplot with panel.xblocks. The result is displayed in Figure 3.3:

- The alternating of years is displayed with blocks of gray and white color using the panel.xblocks function from latticeExtra (line 11). The year is extracted (as character) from the time index of the zoo object with format.POSIXlt (line 3).

- Those values below the mean of each variable are highlighted with short red color blocks at the bottom of each panel, again with the panel.xblocks function (line 15).

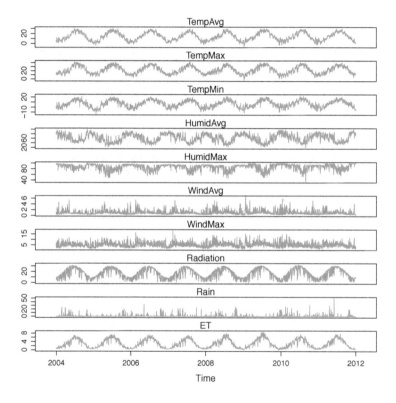

FIGURE 3.1: Time plot of the collection of meteorological time series of the Aranjuez station (`lattice` version).

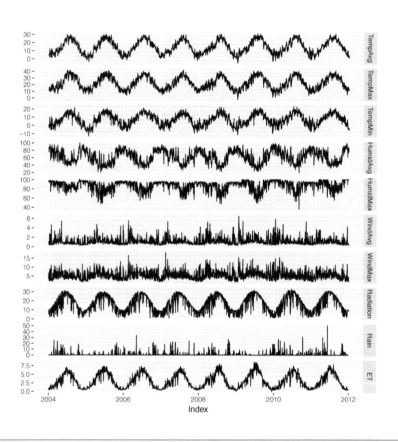

Figure 3.2: Time plot of the collection of meteorological time series of the Aranjuez station (ggplot2 version).

- The label of each time series is displayed with text inside each panel instead of using the strips mechanism. The `panel.text` prints the name of each variable with the aid of `panel.number` (line 21).

- The maxima and minima are highlighted with small blue triangles (lines 25 and 28 respectively).

Because the functions included in the panel function are executed consecutively, their order determines the superposition of graphical layers.

```
1   ## Auxiliary function to extract the year value of a POSIXct time
2   ## index
3   Year <- function(x)format(x, "%Y")
4
5   xyplot(aranjuez,
6          layout = c(1, ncol(aranjuez)),
7          strip = FALSE,
8          scales = list(y = list(cex = 0.6, rot = 0)),
9          panel = function(x, y, ...){
10             ## Alternation of years
11             panel.xblocks(x, Year,
12                           col = c("lightgray", "white"),
13                           border = "darkgray")
14             ## Values under the average highlighted with red regions
15             panel.xblocks(x, y < mean(y, na.rm = TRUE),
16                           col = "indianred1",
17                           height = unit(0.1, 'npc'))
18             ## Time series
19             panel.lines(x, y, col = 'royalblue4', lwd = 0.5, ...)
20             ## Label of each time series
21             panel.text(x[1], min(y, na.rm = TRUE),
22                        names(aranjuez)[panel.number()],
23                        cex = 0.6, adj = c(0, 0), srt = 90, ...)
24             ## Triangles to point the maxima and minima
25             idxMax <- which.max(y)
26             panel.points(x[idxMax], y[idxMax],
27                          col = 'black', fill = 'lightblue', pch = 24)
28             idxMin <- which.min(y)
29             panel.points(x[idxMin], y[idxMin],
30                          col = 'black', fill = 'lightblue', pch = 25)
31          })
```

There is no equivalent `panel.xblocks` function that can be used with ggplot2. Therefore, the ggplot2 version must explicitly compute the corresponding bands (years and regions below the average values):

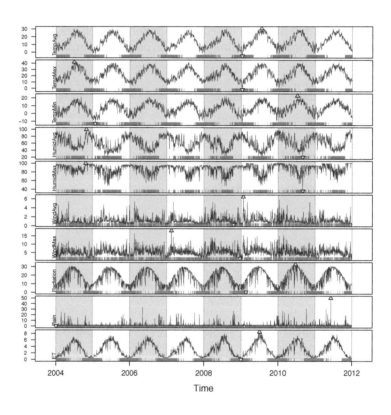

FIGURE 3.3: Enhanced time plot of the collection of meteorological time series of the Aranjuez station.

- The first step in working with ggplot is to transform the zoo object into a data.frame in long format. fortify returns a data.frame with three columns: the time Index, a factor indicating the Series, and the corresponding Value.

```
timeIdx <- index(aranjuez)

aranjuezLong <- fortify(aranjuez, melt = TRUE)

summary(aranjuezLong)
```

```
      Index                 Series            Value
Min.    :2004-01-01   TempAvg  : 2898   Min.    :-12.98
1st Qu.:2005-12-29   TempMax  : 2898   1st Qu.:  2.01
Median :2008-01-09   TempMin  : 2898   Median :  8.62
Mean    :2008-01-03   HumidAvg : 2898   Mean    : 22.05
3rd Qu.:2010-01-03   HumidMax: 2898    3rd Qu.: 27.62
Max.    :2011-12-31   WindAvg  : 2898   Max.    :100.00
                      (Other) :11592   NA's    :37
```

- The bands of values below the average can be easily extracted with scale because these regions are negative when the data.frame is centered.

```
## Values below mean are negative after being centered
scaled <- fortify(scale(aranjuez, scale = FALSE), melt = TRUE)
## The 'scaled' column is the result of the centering.
## The new 'Value' column store the original values.
scaled <- transform(scaled, scaled = Value,
                     Value = aranjuezLong$Value)
underIdx <- which(scaled$scaled <= 0)
## 'under' is the subset of values below the average
under <- scaled[underIdx,]
```

- The years bands are defined with the function endpoints from the xts package:

```
library(xts)
ep <- endpoints(timeIdx, on = 'years')
ep <- ep[-1]
N <- length(ep)
## 'tsp' is start and 'tep' is the end of each band. One of each
    two
## years are selected.
tep <- timeIdx[ep[seq(1, N, 2)] + 1]
tsp <- timeIdx[ep[seq(2, N, 2)]]
```

- The minima and maxima points of each variable are extracted with apply:

```
minIdx <- timeIdx[apply(aranjuez, 2, which.min)]
minVals <- apply(aranjuez, 2, min, na.rm = TRUE)
mins <- data.frame(Index = minIdx,
                   Value = minVals,
                   Series = names(aranjuez))

maxIdx <- timeIdx[apply(aranjuez, 2, which.max)]
maxVals <- apply(aranjuez, 2, max, na.rm = TRUE)
maxs <- data.frame(Index = maxIdx,
                   Value = maxVals,
                   Series = names(aranjuez))
```

- With ggplot we define the canvas, and the layers of information are added successively:

```
ggplot(data = aranjuezLong, aes(Index, Value)) +
    ## Time series of each variable
    geom_line(colour = "royalblue4", lwd = 0.5) +
    ## Year bands
    annotate('rect',
             xmin = tsp, xmax = tep,
             ymin = -Inf, ymax = Inf,
             alpha = 0.4) +
    ## Values below average
    geom_rug(data = under,
             sides = 'b', col = 'indianred1') +
    ## Minima
    geom_point(data = mins, pch = 25) +
```

```
## Maxima
geom_point(data = maxs, pch = 24) +
## Axis labels and theme definition
labs(x = 'Time', y = NULL) +
theme_bw() +
## Each series is displayed in a different panel with an
## independent y scale
facet_free()
```

Some messages from Figure 3.3:

- The radiation, temperature, and evotranspiration are quasi-periodic and are almost synchronized between them. Their local maxima appear in the summer and the local minima in the winter. Obviously, the summer values are higher than the average.

- The average humidity varies in oposition to the temperature and radiation cycle, with local maxima located during winter.

- The average and maximum wind speed, and rainfall vary in a more erratic way and do not show the evident periodic behavior of the radiation and temperature.

- The rainfall is different from year to year. The remaining variables do not show variations between years.

- The fluctuations of solar radiation are more apparent than the temperature fluctuations. There is hardly any day with temperatures below the average value during summer, while it is not difficult to find days with radiation below the average during this season.

3.2 Time Series of Variables with the Same Scale

As an example of time series of variables with the same scale, we will use measurements of solar radiation from different meteorological stations.

The first attempt to display this multivariate time series makes use of the xyplot.zoo method. The objective of this graphic is to display the behavior of the collection as a whole: the series are superposed in the same panel (superpose=TRUE) without legend (auto.key=FALSE), using thin lines and partial transparency[1]. Transparency softens overplotting

---

[1] A similar result can be obtained with autoplot using facets=NULL.

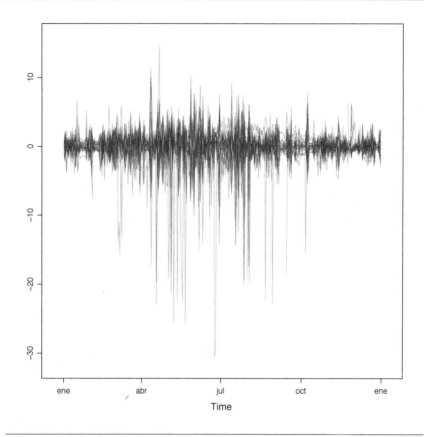

FIGURE 3.4: Time plot of the variations around time average of solar radiation measurements from the meteorological stations of Navarra.

problems and reveals density clusters because regions with more overlapping lines are darker. Figure 3.4 displays the variations around the time average (avRad).

```
load('data/navarra.RData')

avRad <- zoo(rowMeans(navarra, na.rm = 1), index(navarra))
pNavarra <- xyplot(navarra - avRad,
                   superpose = TRUE, auto.key = FALSE,
                   lwd = 0.5, alpha = 0.3, col = 'midnightblue')
pNavarra
```

This result can be improved with different methods: the cut-and-stack method, and the horizon graph with `horizonplot`.

3.2.1 Aspect Ratio and Rate of Change

When a graphic is intended to inform about the rate of change, special attention must be paid to the aspect ratio of the graph, defined as the ratio of the height to the width of the graphical window. Cleveland analyzed the importance of the aspect ratio for judging rate of change (Cleveland and McGill 1984). He concluded that we visually decode the information about the relative local rate of change of one variable with another by comparing the orientations of the local line segments that compose the polylines. The recommendation is to choose the aspect ratio so that the absolute values of the orientations of the segments are centered on 45° (banking to 45°).

The problem with banking to 45° is that the resulting aspect ratio is frequently too small. A suitable solution to minimize wasted space is the cut-and-stack method. The `xyplot.ts` method implements this solution with the combination of the arguments `aspect` and `cut`. The version of Figure 3.4 using banking to 45° and the cut-and-stack method is produced with (Figure 3.5):

```
xyplot(navarra - avRad,
       aspect = 'xy', cut = list(n = 3, overlap = 0.1),
       strip = FALSE,
       superpose = TRUE, auto.key = FALSE,
       lwd = 0.5, alpha = 0.3, col = 'midnightblue')
```

3.2.2 The Horizon Graph

The horizon graph is useful in examining how a large number of series changes over time, and does so in a way that allows both comparisons between the individual time series and independent analysis of each series. Moreover, extraordinary behaviors and predominant patterns are easily distinguished (Heer, Kong, and Agrawala 2009; Few 2008).

This graph displays several stacked series collapsing the y-axis to free vertical space:

- Positive and negative values share the same vertical space. Negative values are inverted and placed above the reference line. Sign is encoded using different hues (positive values in blue and negative values in red).

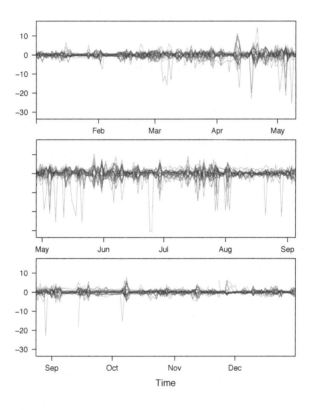

FIGURE 3.5: Cut-and-stack plot with banking to 45°.

- Differences in magnitude are displayed as differences in color intensity (darker colors for greater differences).

- The color bands share the same baseline and are superposed, with darker bands in front of the lighter ones.

Because the panels share the same design structure, once this technique is understood, it is easy to establish comparisons or spot extraordinary events. This method is what Tufte described as small multiples (Tufte 1990).

Figure 3.6 displays the variations of solar radiation around the time average with an horizon graph using a row for each time series. In the code we choose `origin=0` and leave the argument `horizonscale` undefined (default). With this combination each panel has different scales and the colors in each panel represent deviations from the origin. This is depicted in the color key with the Δ_i symbol, where the subscript i denotes the existence of multiple panels with different scales.

```
library(latticeExtra)

horizonplot(navarra - avRad,
            layout = c(1, ncol(navarra)),
            origin = 0, ## Deviations in each panel are calculated
                      ## from this value
            colorkey = TRUE)
```

Figure 3.6 allows several questions to be answered:

- Which stations consistently measure above and below the average?

- Which stations resemble more closely the average time series?

- Which stations show erratic and uniform behavior?

- In each of the stations, is there any day with extraordinary measurements?

- Which part of the year is associated with more intense absolute fluctuations across the set of stations?

3.2.3 Time Graph of the Differences between a Time Series and a Reference

The horizon graph is also useful in revealing the differences between a univariate time series and another reference. For example, we might be interested in the departure of the observed temperature from the long-term average, or in other words, the temperature change over time.

Let's illustrate this approach with the time series of daily average temperatures measured at the meteorological station of Aranjuez. The reference is the long-term daily average calculated with ave.

```
Ta <- aranjuez$TempAvg
timeIndex <- index(aranjuez)
longTa <- ave(Ta, format(timeIndex, '%j'))
diffTa <- (Ta - longTa)
```

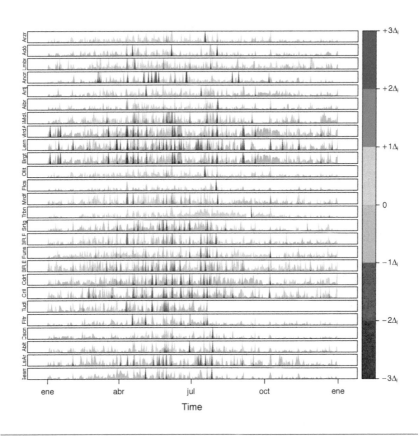

FIGURE 3.6: Horizon plot of variations around time average of solar radiation measurements from the meteorological stations of Navarra. The Δ_i symbol in the color key represents the deviation in each panel from the origin value.

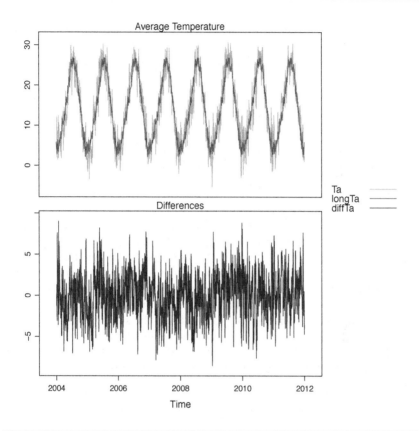

FIGURE 3.7: Daily temperature time series, its long-term average and the differences between them.

The temperature time series, the long-term average and the differences between them can be displayed with the xyplot method, now using screens to use a different panel for the differences time series (Figure 3.7)

```
xyplot(cbind(Ta, longTa, diffTa),
    col = c('darkgray', 'red', 'midnightblue'),
    superpose = TRUE, auto.key = list(space = 'right'),
    screens = c(rep('Average Temperature', 2), 'Differences'))
```

The horizon graph is better suited for displaying the differences. The next code again uses the cut-and-stack method (Figure 3.5) to distinguish between years. Figure 3.8 shows that 2004 started clearly above the average while 2005 and 2009 did the contrary. Year 2007 was frequently below the long-term average but 2011 was more similar to that reference.

```
years <- unique(format(timeIndex, '%Y'))

horizonplot(diffTa, cut = list(n = 8, overlap = 0),
            colorkey = TRUE, layout = c(1, 8),
            scales = list(draw = FALSE, y = list(relation = 'same'))
            ,
            origin = 0, strip.left = FALSE) +
    layer(grid.text(years[panel.number()], x = 0, y = 0.1,
                    gp = gpar(cex = 0.8),
                    just = "left"))
```

A different approach to display this information is to produce a level plot displaying the time series using parts of its time index as independent and conditioning variables[2]. The following code displays the differences with the day of month on the horizontal axis and the year on the vertical axis, with a different panel for each month number. Therefore, each cell of Figure 3.9 corresponds to a certain day of the time series. If you compare this figure with the horizon plot, you will find the same previous findings but revealed now in more detail. On the other hand, while the horizon plot of Figure 3.8 clearly displays the yearly evolution, the combination of variables of the level plot focuses on the comparison between years in a certain month.

```
year <- function(x)as.numeric(format(x, '%Y'))
day <- function(x)as.numeric(format(x, '%d'))
month <- function(x)as.numeric(format(x, '%m'))

myTheme <- modifyList(custom.theme(region = brewer.pal(9, 'RdBu')
             ),
                  list(
                      strip.background = list(col = 'gray'),
                      panel.background = list(col = 'gray')))

maxZ <- max(abs(diffTa))

levelplot(diffTa ~ day(timeIndex) * year(timeIndex) | factor(
    month(timeIndex)),
```

---

[2]This approach was inspired by the strip function of the metvurst package https://metvurst.wordpress.com/2013/03/04/visualising-large-amounts-of-hourly-environmental-data-2/

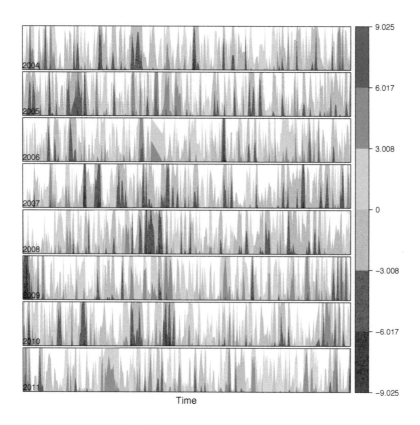

FIGURE 3.8: Horizon graph displaying differences between a daily temperature time series and its long-term average.

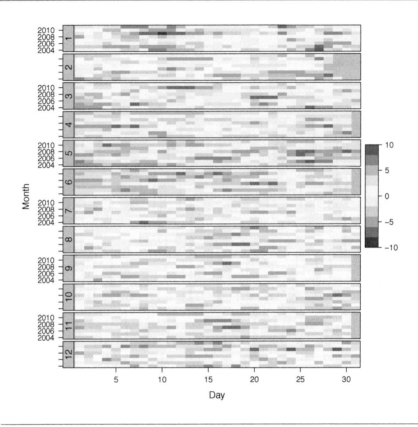

FIGURE 3.9: Level plot of differences between a daily temperature time series and its long-term average.

```
at = pretty(c(-maxZ, maxZ), n = 8),
colorkey = list(height = 0.3),
layout = c(1, 12), strip = FALSE, strip.left = TRUE,
xlab = 'Day', ylab = 'Month',
par.settings = myTheme)
```

The ggplot version of the Figure 3.9 requires a data.frame with the day, year, and month arranged in different columns.

```
df <- data.frame(Vals = diffTa,
                 Day = day(timeIndex),
                 Year = year(timeIndex),
                 Month = month(timeIndex))
```

The values (`Vals` column of this `data.frame`) are displayed as a level plot thanks to the `geom_raster` function.

```
library(scales)
## The packages scales is needed for the pretty_breaks function.

ggplot(data = df,
      aes(fill = Vals,
         x = Day,
         y = Year)) +
   facet_wrap(~ Month, ncol = 1, strip.position = 'left') +
   scale_y_continuous(breaks = pretty_breaks()) +
   scale_fill_distiller(palette = 'RdBu', direction = 1) +
   geom_raster() +
   theme(panel.grid.major = element_blank(),
         panel.grid.minor = element_blank())
```

3.3   Stacked Graphs

If the variables of a multivariate time series can be summed to produce a meaningful global variable, they may be better displayed with stacked graphs. For example, the information on unemployment in the United States provides data of unemployed persons by industry and class of workers, and can be summed to give a total unemployment time series.

```
load('data/unemployUSA.RData')
```

The time series of unemployment can be directly displayed with the `xyplot.zoo` method (Figure 3.10).

```
xyplot(unemployUSA,
      superpose = TRUE,
      par.settings = custom.theme,
      auto.key = list(space = 'right'))
```

This graphical output is not very useful: the legend includes too many items; the vertical scale is dominated by the largest series, with several series buried in the lower part of the scale; the trend, variations and structure of the total and individual contributions cannot be deduced from this graph.

A partial improvement is to display the multivariate time series as a set of stacked colored polygons to follow the macro/micro principle proposed by Tufte (Tufte 1990): Show a collection of individual time series

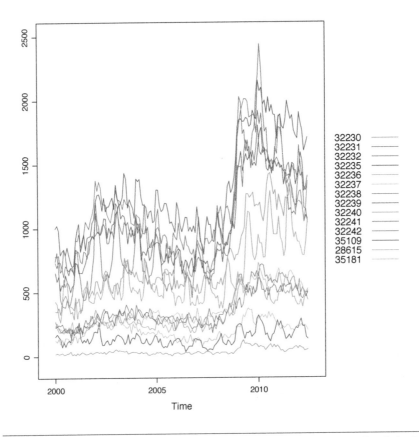

FIGURE 3.10: Time series of unemployment with xyplot using the default panel function.

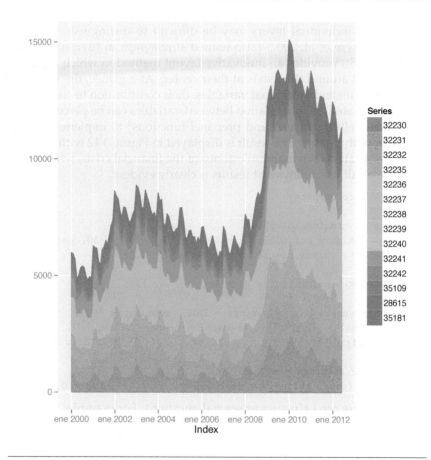

FIGURE 3.11: Time series of unemployment with stacked areas using geom_area.

and also display their sum. A traditional stacked graph is easily obtained with geom_area (Figure 3.11):

```
library(scales) ## scale_x_yearmon needs scales::pretty_breaks
autoplot(unemployUSA, facets = NULL, geom = 'area') +
    geom_area(aes(fill = Series)) +
    scale_x_yearmon()
```

Traditional stacked graphs have their bottom on the x-axis which makes the overall height at each point easy to estimate. On the other hand, with this layout, individual layers may be difficult to distinguish. The *The-meRiver* (Havre et al. 2002) (also named *streamgraph* in (Byron and Wattenberg 2008)) provides an innovative layout method in which layers are symmetrical around the x-axis at their center. At a glance, the pattern of the global sum and individual variables, their contribution to conform the global sum, and the interrelation between variables can be perceived.

I have defined a panel and prepanel functions[3] to implement a The-meRiver with xyplot. The result is displayed in Figure 3.12 with a vertical line to indicate one of main milestones of the financial crisis, whose effect on the overall unemployment results is clearly evident.

```
library(colorspace)
## We will use a qualitative palette from colorspace
nCols <- ncol(unemployUSA)
pal <- rainbow_hcl(nCols, c = 70, l = 75, start = 30, end = 300)
myTheme <- custom.theme(fill = pal, lwd = 0.2)

sep2008 <- as.numeric(as.yearmon('2008-09'))

xyplot(unemployUSA, superpose = TRUE, auto.key = FALSE,
       panel = panel.flow, prepanel = prepanel.flow,
       origin = 'themeRiver', scales = list(y = list(draw = FALSE))
       ,
       par.settings = myTheme) +
    layer(panel.abline(v = sep2008, col = 'gray', lwd = 0.7))
```

This figure can help answer several questions. For example:

- What is the industry or class of worker with the lowest/highest unemployment figures during this time period?

- What is the industry or class of worker with the lowest/highest unemployment increases due to the financial crisis?

- There are a number of local maxima and minima of the total unemployment numbers. Are all the classes contributing to the maxima/minima? Do all the classes exhibit the same fluctuation behavior as the global evolution?

More questions and answers can be found in the "Current Employment Statistics" reports from the Bureau of Labor Statistics[4].

---

[3]The code of these panel and prepanel functions is explained in Section 3.3.1.

[4]The March 2012 highlights report is available at http://www.bls.gov/ces/highlights032012.pdf.

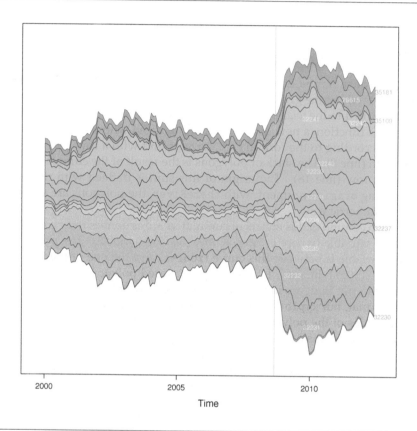

FIGURE 3.12: ThemeRiver of unemployment in the United States.

3.3.1 ✍Panel and Prepanel Functions to Implement the ThemeRiver with `xyplot`

The `xyplot` function displays information according to the class of its first argument (methods) and to the `panel` function. We will use the `xyplot.zoo` method (equivalent to the `xyplot.ts` method) with a new custom `panel` function. This new panel function has four main arguments, three of them calculated by `xyplot` (x, y and groups) and a new one, `origin`. Of course, it includes the ... argument to provide additional arguments.

The first step is to create a `data.frame` with coordinates and with the `groups` factor (line 3). The value and number of the levels will be used in the main step of this panel function. With this data.frame we have to calculate the y and x coordinates for each group to get a stacked set of polygons.

This data.frame is in the *long* format, with a row for each observation, and where the group column identifies the variable. Thus, it must be transformed to the *wide* format, with a column for each variable. With the unstack function, a new data.frame is produced whose columns are defined according to the formula y ~ groups and with a row for each time position (line 8). The stack of polygons is the result of the cumulative sum of each row (line 18). The origin of this sum is defined with the corresponding origin argument: with themeRiver, the polygons are arranged in a symmetric way.

Each column of this matrix of cumulative sums defines the y coordinate of each variable (where origin is now the first variable). The polygon of each variable is between this curve (iCol+1) and the one of the previous variable (iCol) (line 19). In order to get a closed polygon, the coordinates of the inferior limit are in reverse order. This new data.frame (Y) is in the *wide* format, but xyplot requires the information in the *long* format: the y coordinates of the polygons are extracted from the values column of the *long* version of this data.frame (line 24).

The x coordinates are produced in an easier way. Again, unstack produces a data.frame with a column for each variable and a row for each time position (line 27), but now, because the x coordinates are the same for the set of polygons, the corresponding vector is constructed directly using a combination of concatenation and repetition (line 28).

Finally, the groups vector is produced, repeating each element of the columns of the original data.frame (dat$groups) twice to account for the forward and reverse curves of the corresponding polygon (line 30).

The final step before displaying the polygons is to acquire the graphical settings. The information retrieved with trellis.par.get is transferred to the corresponding arguments of panel.polygon (line 33).

Everything is ready for constructing the polygons. With a for loop (line 39), the coordinates of the corresponding group are extracted from the x and y vectors, and a polygon is displayed with panel.polygon. The labels of each polygon (the levels of the original groups variable, groupLevels) are printed inside the polygon if there is enough room for the text (hChar>1) or at the right if the polygon is too small, or if it is the first or last variable of the set (line 58). Both the polygons and the labels share the same color (col[i]).

```
1  panel.flow <- function(x, y, groups, origin, ...)
```

```
2    {
3       dat <- data.frame(x = x, y = y, groups = groups)
4       nVars <- nlevels(groups)
5       groupLevels <- levels(groups)
6
7       ## From long to wide
8       yWide <- unstack(dat, y~groups)
9       ## Where are the maxima of each variable located? We will use
10      ## them to position labels.
11      idxMaxes <- apply(yWide, 2, which.max)
12
13      ##Origin calculated following Havr.eHetzler.ea2002
14      if (origin=='themeRiver') origin = -1/2*rowSums(yWide)
15      else origin = 0
16      yWide <- cbind(origin = origin, yWide)
17      ## Cumulative sums to define the polygon
18      yCumSum <- t(apply(yWide, 1, cumsum))
19      Y <- as.data.frame(sapply(seq_len(nVars),
20                         function(iCol)c(yCumSum[,iCol+1],
21                                         rev(yCumSum[,iCol]))))
22      names(Y) <- levels(groups)
23      ## Back to long format, since xyplot works that way
24      y <- stack(Y)$values
25
26      ## Similar but easier for x
27      xWide <- unstack(dat, x~groups)
28      x <- rep(c(xWide[,1], rev(xWide[,1])), nVars)
29      ## Groups repeated twice (upper and lower limits of the
30         polygon)
31      groups <- rep(groups, each = 2)
32
33      ## Graphical parameters
34      superpose.polygon <- trellis.par.get("superpose.polygon")
35      col = superpose.polygon$col
36      border = superpose.polygon$border
37      lwd = superpose.polygon$lwd
38
39      ## Draw polygons
40      for (i in seq_len(nVars)){
41          xi <- x[groups==groupLevels[i]]
42          yi <- y[groups==groupLevels[i]]
43          panel.polygon(xi, yi, border = border,
44                        lwd = lwd, col = col[i])
45      }
```

```
45
46      ## Print labels
47      for (i in seq_len(nVars)){
48          xi <- x[groups==groupLevels[i]]
49          yi <- y[groups==groupLevels[i]]
50          N <- length(xi)/2
51          ## Height available for the label
52          h <- unit(yi[idxMaxes[i]], 'native') -
53              unit(yi[idxMaxes[i] + 2*(N-idxMaxes[i]) +1], 'native')
54          ##...converted to "char" units
55          hChar <- convertHeight(h, 'char', TRUE)
56          ## If there is enough space and we are not at the first or
57          ## last variable, then the label is printed inside the
                polygon.
58          if((hChar >= 1) && !(i %in% c(1, nVars))){
59              grid.text(groupLevels[i],
60                      xi[idxMaxes[i]],
61                      (yi[idxMaxes[i]] +
62                       yi[idxMaxes[i] + 2*(N-idxMaxes[i]) +1])/2,
63                      gp = gpar(col = 'white', alpha = 0.7, cex = 0.7)
                        ,
64                      default.units = 'native')
65          } else {
66              ## Elsewhere, the label is printed outside
67
68              grid.text(groupLevels[i],
69                      xi[N],
70                      (yi[N] + yi[N+1])/2,
71                      gp = gpar(col = col[i], cex = 0.7),
72                      just = 'left', default.units = 'native')
73          }
74      }
75  }
```

With this panel function, xyplot displays a set of stacked polygons corresponding to the multivariate time series (Figure 3.13). However, the graphical window is not large enough, and part of the polygons fall out of it. Why?

```
xyplot(unemployUSA, superpose = TRUE, auto.key = FALSE,
       panel = panel.flow, origin = 'themeRiver',
       par.settings = myTheme, cex = 0.4, offset = 0,
       scales = list(y = list(draw = FALSE)))
```

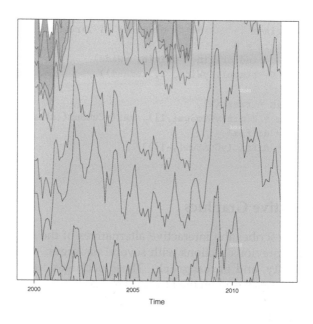

FIGURE 3.13: First attempt of ThemeRiver.

The problem is that lattice makes a preliminary estimate of the window size using a default prepanel function that is unaware of the internal calculations of our new panel.flow function. The solution is to define a new prepanel.flow function.

The input arguments and first lines are the same as in panel.flow. The output is a list whose elements are the limits for each axis (xlim and ylim, line 12), and the sequence of differences (dx and dy, line 14) that can be used for the aspect and banking calculations.

The limits of the x-axis are defined with the range of the time index, while the limits of the y-axis are calculated with the minimum of the first column of yCumSum (the origin line) and with the maximum of its last column (the upper line of the cumulative sum) (line 13).

```
1  prepanel.flow <- function(x, y, groups, origin,...)
2  {
3      dat <- data.frame(x = x, y = y, groups = groups)
4      nVars <- nlevels(groups)
5      groupLevels <- levels(groups)
6      yWide <- unstack(dat, y~groups)
```

```
7    if (origin=='themeRiver') origin = -1/2*rowSums(yWide)
8    else origin = 0
9    yWide <- cbind(origin = origin, yWide)
10   yCumSum <- t(apply(yWide, 1, cumsum))
11
12   list(xlim = range(x),
13        ylim = c(min(yCumSum[,1]), max(yCumSum[,nVars+1])),
14        dx = diff(x),
15        dy = diff(c(yCumSum[,-1])))
16 }
```

3.4 Interactive Graphics

This section describes the interactive alternatives of the static figures included in the previous sections with several packages: dygraphs, highcharter, plotly, and streamgraph. These packages are R interfaces to JavaScript libraries based on the htmlwidgets package.

3.4.1 Dygraphs

The dygraphs package is an interface to the dygraphs JavaScript library, and provides facilities for charting time-series. It works automatically with xts time series objects, or with objects than can be coerced to this class. The result is an interactive graph, where values are displayed according to the mouse position over the time series. Regions can be selected to zoom into a time period. Figure 3.14 is an snapshot of the interactive graph.

```
library(dygraphs)

dyTemp <- dygraph(aranjuez[, c("TempMin", "TempAvg", "TempMax")],
                  main = "Temperature in Aranjuez",
                  ylab = "°C")

dyTemp
```

You can customize dygraphs by piping additional commands onto the original graphic. The function dyOptions provides several choices for the graphic, and the function dyHighlight configures options for data series mouse-over highlighting. For example, with the next code the semi-transparency value of the non-selected lines is reduced and the width of selected line is increased (Figure 3.15).

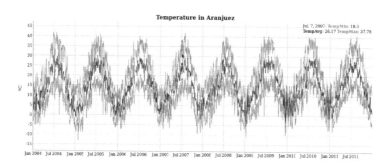

FIGURE 3.14: Snapshot of an interactive graphic produced with dygraphs.

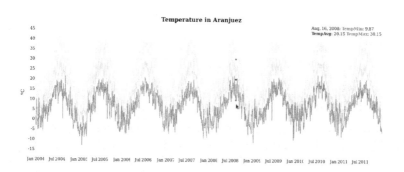

FIGURE 3.15: Snapshot of a selection in an interactive graphic produced with dygraphs.

```
dyTemp %>%
    dyHighlight(highlightSeriesBackgroundAlpha = 0.2,
                highlightSeriesOpts = list(strokeWidth = 2))
```

An alternative approach to depict the upper and lower variables of this time series is with a shaded region. The dySeries function accepts a character vector of length 3 that specifies a set of input column names to use as the lower, value, and upper for a series with a shaded region around it (Figure 3.16).

```
dygraph(aranjuez[, c("TempMin", "TempAvg", "TempMax")], .
```

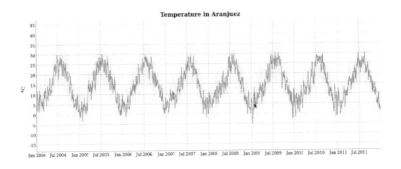

FIGURE 3.16: Shaded region between upper and lower values around a time series.

```
       main = "Temperature in Aranjuez",
       ylab = "ºC") %>%
dySeries(c("TempMin", "TempAvg", "TempMax"),
       label = "Temperature")
```

3.4.2 Highcharter

The highcharter package is an interface to the highcharts JavaScript library, with a wide spectrum of graphics solutions. Displaying time series with this package can be achieved with the combination of the generic highchart function and several calls to the hc_add_series_xts function through the pipe %>% operator. Once again, the result is an interactive graph with selection and zoom capabilities. Figure 3.17 is an snapshot of the interactive graph, and Figure 3.18 is an snapshot of this same graph with zoom.

```
library(highcharter)
library(xts)

aranjuezXTS <- as.xts(aranjuez)

highchart() %>%
   hc_add_series(name = 'TempMax',
                 aranjuezXTS[, "TempMax"]) %>%
   hc_add_series(name = 'TempMin',
                 aranjuezXTS[, "TempMin"]) %>%
```

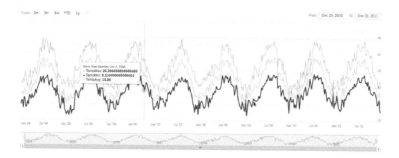

FIGURE 3.17: Snapshot of an interactive graphic produced with `high-charter`.

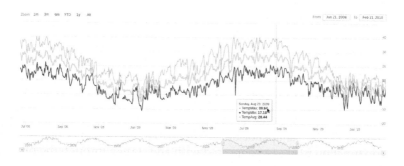

FIGURE 3.18: Snapshot of a zoom in an interactive graphic produced with `highcharter`.

```
hc_add_series(name = 'TempAvg',
              aranjuezXTS[, "TempAvg"])
```

3.4.3 plotly

The `plotly` package is an interface to the `plotly` JavaScript library, also with a wide spectrum of graphics solutions. This package does not provide any function specifically focused on time series. Thus, the time series object has to be transformed in a `data.frame` including a column for the time index. If the `data.frame` is in *wide* format (one column per variable), each variable will be represented with a call to the `add_lines` function. However, if the `data.frame` is in *long* format (a column for values, and a column for variable names) only one call to `add_lines` is required. The next code follows this approach using the combination of `fortify`, to convert the `zoo` object into a `data.frame`, and `melt`, to transform from wide to long format.

```
aranjuezDF <- fortify(aranjuez[,
                       c("TempMax",
                         "TempAvg",
                         "TempMin")],
            melt = TRUE)


summary(aranjuezDF)
```

Index	Series	Value
Min. :2004-01-01	TempMax:2898	Min. :-12.980
1st Qu.:2005-12-29	TempAvg:2898	1st Qu.: 7.107
Median :2008-01-09	TempMin:2898	Median : 13.560
Mean :2008-01-03		Mean : 14.617
3rd Qu.:2010-01-03		3rd Qu.: 21.670
Max. :2011-12-31		Max. : 41.910
		NA's :10

Figure 3.19 is a snapshot of the interactive graphic produce with the generic function `plot_ly` connected with `add_lines` through the pipe operator, `%>%`.

```
library(plotly)


plot_ly(aranjuezDF) %>%
    add_lines(x = ~ Index,
              y = ~ Value,
              color = ~ Series)
```

3.4.4 streamgraph

The `streamgraph` package[5] creates interactive stream graphs based on the `htmlwidgets` package and the `D3.js` JavaScript library. Its main function, `streamgraph`, requires a `data.frame` as the first argument. Besides, its

---

[5]The streamgraph package, http://hrbrmstr.github.io/streamgraph/, is not available in CRAN. It can be installed using the devtools or the remotes package.

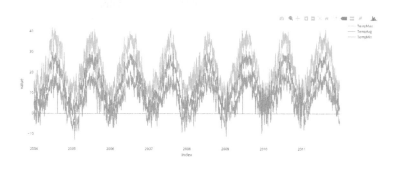

FIGURE 3.19: Snapshot of an interactive graphic produced with `plotly`.

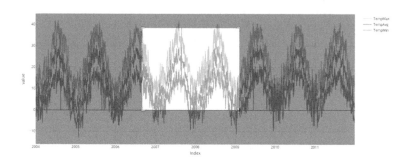

FIGURE 3.20: Snapshot of a zoom in an interactive graphic produced with `plotly`.

three next arguments, key, value, and date, make this function a good candidate to work together with fortify and melt.

```
unemployDF <- fortify(unemployUSA, melt = TRUE)

head(unemployDF)
```

```
    Index Series Value
1 ene 2000  32230    19
2 feb 2000  32230    25
3 mar 2000  32230    17
4 abr 2000  32230    20
5 may 2000  32230    27
6 jun 2000  32230    13
```

Figures 3.21 and 3.22 are snapshots of the interactive graphic created with the functions streamgraph, sg_axis, and sg_fill_brewer, connected through the pipe operator, %>%.

```
## remotes::install_github("hrbrmstr/streamgraph")
library(streamgraph)

streamgraph(unemployDF,
            key = "Series",
            value = "Value",
            date = "Index") %>%
      sg_axis_x(1, "year", "%Y") %>%
      sg_fill_brewer("Set1")
```

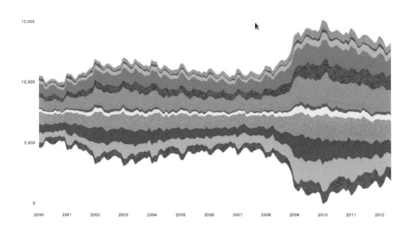

FIGURE 3.21: Streamgraph created with the streamgraph package, without selection.

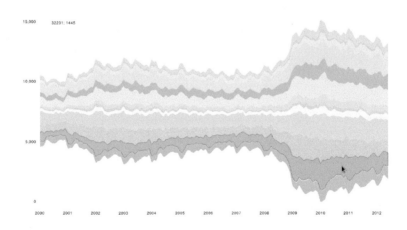

FIGURE 3.22: Streamgraph created with the streamgraph package, with a selection.

Chapter 4

Time as a Conditioning or Grouping Variable

In Section 3.1 we learned to display the time evolution of multiple time series with different scales. But, what if instead of displaying the time evolution we want to represent the relation between the variables? This chapter follows this approach: Section 4.1 proposes the scatterplot matrix solution using groups defined according to the time as a grouping variable; Section 4.2 produces an enhanced scatterplot with time as a conditioning variable using the small multiples technique; Section 4.1.1 includes a discussion about the hexagonal binning for large datasets.

Along this chapter these subjects are covered: scatterplot matrices, small multiples, hexagonal binning, and panel functions.

The most relevant packages in this chapter are: zoo for reading and arranging data as time series; GGally for creating the scatterplot matrix with ggplot2; hexbin for hexagonal binning; reshape2 for converting data from the wide format to the long format.

4.1 Scatterplot Matrix: Time as a Grouping Variable

The scatterplot matrices are based on the technique of small multiples (Tufte 1990): small, thumbnail-sized representations of multiple images displayed all at once, which allows the reader to immediately, and in parallel, compare the inter-frame differences. A scatterplot matrix is a display of all pairwise bivariate scatterplots arranged in a $p \times p$ matrix for p variables. Each subplot shows the relation between the pair of variables at the intersection of the row and column indicated by the variable names in the diagonal panels (Friendly and Denis 2005).

This graphical tool is implemented in the splom function. The following code displays the relation between the set of meteorological variables using a sequential palette from the ColorBrewer catalog (RbBu, with black added to complete a twelve-color palette) to encode the month. The order of colors of this palette is chosen in order to display summer months with intense colors and to distinguish between the first and second half of the the year with red and blue, respectively (Figure 4.1).

```
library(zoo)

load('data/aranjuez.RData')
aranjuezDF <- as.data.frame(aranjuez)
aranjuezDF$Month <- format(index(aranjuez), '%m')

## Red-Blue palette with black added (12 colors)
colors <- c(brewer.pal(n = 11, 'RdBu'), '#000000')
## Rearrange according to months (darkest for summer)
colors <- colors[c(6:1, 12:7)]

splom(~ aranjuezDF,
      groups = aranjuezDF$Month,
      auto.key = list(space = 'right',
                  title = 'Month', cex.title = 1),
      pscale = 0, varname.cex = 0.7, xlab = '',
      par.settings = custom.theme(symbol = colors,
                          pch = 19),
      cex = 0.3, alpha = 0.1)
```

A bit of interactivity can be added to this plot with the identification of some points. This task is easy with panel.link.splom. The points are selected via mouse clicks (and highlighted in green). Clicks other than left-clicks terminate the procedure. The output of this function is the index of chosen points.

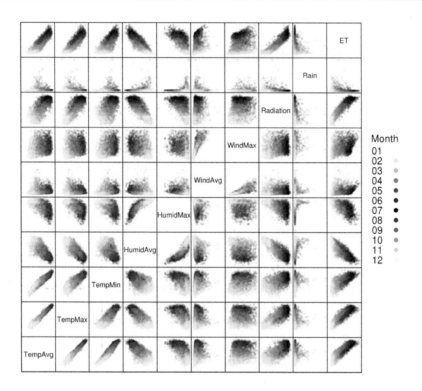

FIGURE 4.1: Scatter plot matrix of the collection of meteorological time series of the Aranjuez station.

```
trellis.focus('panel', 1, 1)
idx <- panel.link.splom(pch = 13, cex = 0.6, col = 'green')
aranjuez[idx,]
```

The ggplot2 version of Figure 4.1 is produced thanks to the ggpairs function provided by the GGally package.

```
library(GGally)

ggpairs(aranjuezDF,
        columns = 1:10, ## Do not include "Month"
        upper = list(continuous = "points"),
        mapping = aes(colour = Month, alpha = 0.1))
```

Let's explore Figure 4.1. For example,

- The highest values of ambient temperature (average, maximum, and minimum), solar radiation, and evapotranspiration can be found during the summer.

- These variables are almost linearly related. The relation between radiation and temperature is different during both halves of the year (red and blue regions can be easily distinguished).

- The humidity reaches its highest values during winter without appreciable differences between the first and second half of the year. The temperature and humidity may be related with an exponential function.

4.1.1 Hexagonal Binning

For large datasets, the display of a large number of points in a scatterplot produces hidden point density, long computation times, and slow displays. These problems can be circumvented with the estimation and representation of points densities. A common encoding uses gray scales, pseudo colors or partial transparency. An improved scheme encodes density as the size of hexagon symbols inscribed within hexagonal binning regions (D. B. Carr et al. 1987).

The hexbin package (D. Carr, Lewin-Koh, and Maechler 2018) includes several functions for hexagonal binning. The panel.hexbinplot is a good substitute for the default panel function. In addition, our first attempt with splom can be improved with several modifications (Figure 4.2):

- The panels of the lower part of the matrix (lower.panel) will include a locally weighted scatterplot smoothing (loess) with panel.loess (line 12).

- The diagonal panels (diag.panel) will display the kernel density estimate of each variable (line 5). The density function computes this estimate. The result is adjusted to the panel limits (calculated with current.panel.limits). The kernel density is plotted with panel.lines and the diag.panel.splom function completes the content of each diagonal panel.

- The scale's ticks and labels are suppressed with pscale=0 (line 17)

- The point density is encoded with the default palette, LinGray, (darker colors for high density values and lighter colors for almost empty regions, with a gradient of grays for intermediate values).

```
1   library(hexbin)
2
3   splom(~as.data.frame(aranjuez),
4       panel = panel.hexbinplot,
5       diag.panel = function(x, ...){
6           yrng <- current.panel.limits()$ylim
7           d <- density(x, na.rm = TRUE)
8           d$y <- with(d, yrng[1] + 0.95 * diff(yrng) * y / max(y))
9           panel.lines(d)
10          diag.panel.splom(x, ...)
11      },
12      lower.panel = function(x, y, ...){
13          panel.hexbinplot(x, y, ...)
14          panel.loess(x, y, ..., col = 'red')
15      },
16      xlab = '',
17      pscale = 0,
18      varname.cex = 0.7)
```

A drawback of the matrix of scatterplots with hexagonal binning is that each panel is drawn independently, so it is impossible to compute a common color key for all of them. In other words, two cells with exactly the same color in different panels encode different point densities.

It is possible to display a reduced set of variables against another one and generate a common color key using the hexbinplot function. First, the dataset must be reshaped from the wide format (one colum for each variable) to the long format (only one column for the temperature values with one row for each observation). This task is easily accomplished with the melt function included in the reshape2 package.

```
library(reshape2)

aranjuezRshp <- melt(aranjuezDF,
                measure.vars = c('TempMax',
                                 'TempAvg',
                                 'TempMin'),
                variable.name = 'Statistic',
                value.name = 'Temperature')

summary(aranjuezRshp)

   HumidAvg HumidMax WindAvg WindMax
  Min. :19.89 Min. : 35.88 Min. :0.250 Min. : 1.550
```

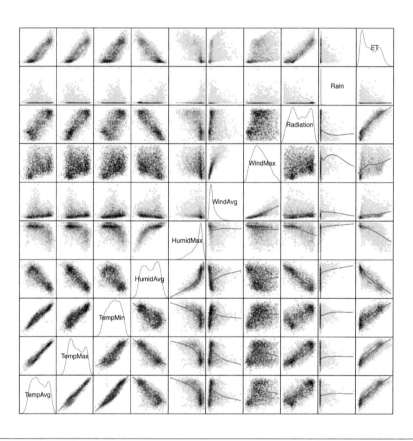

FIGURE 4.2: Scatterplot matrix of the collection of meteorological time series of the Aranjuez station using hexagonal binning.

```
1st Qu.:47.04 1st Qu.: 81.60 1st Qu.:0.670 1st Qu.: 3.780
Median :62.49 Median : 90.90 Median :0.920 Median : 5.030
Mean :62.11 Mean : 87.20 Mean :1.166 Mean : 5.216
3rd Qu.:77.30 3rd Qu.: 94.90 3rd Qu.:1.430 3rd Qu.: 6.540
Max. :99.50 Max. :100.00 Max. :6.450 Max. :10.000
NA's :6 NA's :33 NA's :345
  Radiation Rain ET Month
Min. : 0.28 Min. : 0.000 Min. :0.000 Length:8694
1st Qu.: 9.37 1st Qu.: 0.000 1st Qu.:1.160 Class :character
Median :16.67 Median : 0.000 Median :2.750 Mode :character
Mean :16.73 Mean : 1.046 Mean :3.088
3rd Qu.:24.63 3rd Qu.: 0.200 3rd Qu.:4.923
Max. :32.74 Max. :49.730 Max. :8.560
                       NA's :42
  Statistic Temperature
TempMax:2898 Min. :-12.980
TempAvg:2898 1st Qu.: 7.107
TempMin:2898 Median : 13.560
             Mean : 14.617
             3rd Qu.: 21.670
             Max. : 41.910
             NA's :10
```

The hexbinplot displays this dataset with a different panel for each type of temperature (average, maximum, and minimum) but with a common color key encoding the point density (Figure 4.3). Now, two cells with the same color in different panels encode the same value.

```
hexbinplot(Radiation ~ Temperature | Statistic,
          data = aranjuezRshp,
          layout = c(1, 3)) +
    layer(panel.loess(..., col = 'red'))
```

The ggplot2 version is based on the stat_binhex function.

```
ggplot(data = aranjuezRshp,
      aes(Temperature, Radiation)) +
    stat_binhex(ncol = 1) +
    stat_smooth(se = FALSE, method = 'loess', col = 'red') +
    facet_wrap(~ Statistic, ncol = 1) +
    theme_bw()
```

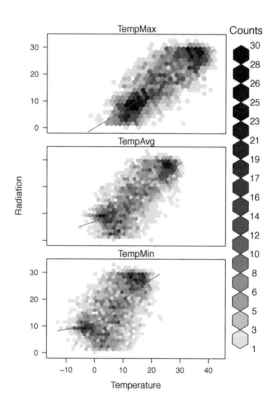

FIGURE 4.3: Scatterplot with hexagonal binning of temperature versus solar radiation using data of the Aranjuez station (`lattice` version).

4.2 Scatterplot with Time as a Conditioning Variable

After discussing the hexagonal binning, let's recover the time variable. Figure 4.1 uses colors to encode months. Instead, we will now display separate scatterplots with a panel for each month. In addition, the statistic type (average, maximum, minimum) is included as an additional conditioning variable.

This matrix of panels can be displayed with ggplot using facet_grid. The code of Figure 4.4 uses partial transparency to cope with overplotting, small horizontal and vertical segments (geom_rug) to display points density on both variables, and a smooth line in each panel.

```
ggplot(data = aranjuezRshp, aes(Radiation, Temperature)) +
    facet_grid(Statistic ~ Month) +
    geom_point(col = 'skyblue4', pch = 19, cex = 0.5, alpha = 0.3)
        +
    geom_rug() +
    stat_smooth(se = FALSE, method = 'loess',
                col = 'indianred1', lwd = 1.2) +
    theme_bw()
```

The version with lattice needs the useOuterStrips function from the latticeExtra package, which prints the names of the conditioning variables on the top and left outer margins (Figure 4.5).

```
useOuterStrips(
    xyplot(Temperature ~ Radiation | Month * Statistic,
        data = aranjuezRshp,
        between = list(x = 0),
        col = 'skyblue4', pch = 19,
        cex = 0.5, alpha = 0.3)) +
    layer({
        panel.rug(..., col.line = 'indianred1',
                end = 0.05, alpha = 0.6)
        panel.loess(..., col = 'indianred1',
                lwd = 1.5, alpha = 1)
    })
```

These figures show the typical seasonal behavior of solar radiation and ambient temperature. Additionally, it displays in more detail the same relations between radiation and temperature already discussed with Figure 4.3.

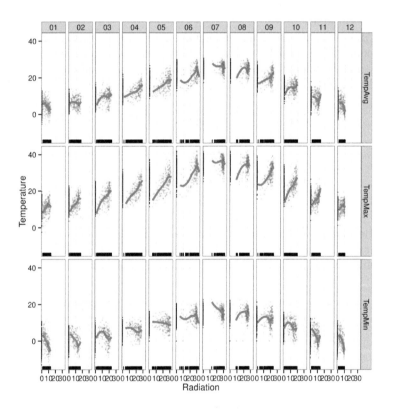

FIGURE 4.4: Scatterplot of temperature versus solar radiation for each month using data of the Aranjuez station (ggplot2 version).

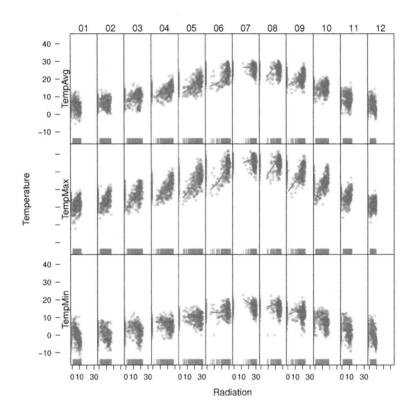

FIGURE 4.5: Scatterplot of temperature versus solar radiation for each month using data of the Aranjuez station (lattice version).

Chapter 5

Time as a Complementary Variable

In this chapter, time will be used as a complementary variable which adds information to a graph where several variables are confronted. We will illustrate this approach with the evolution of the relationship between Gross National Income (GNI) and carbon dioxide (CO_2) emissions for a set of countries extracted from the database of the World Bank Open Data. We will try several solutions to display the relationship between CO_2 emissions and GNI over the years using time as a complementary variable.

Along this chapter these subjects are covered: qualitative palettes, visual discrimination, panel functions, label positioning, small multiples, class intervals, and animation. Last section is devoted to interactive graphics.

The most relevant packages in this chapter are: zoo for reading and arranging data as time series; reshape2 for converting data from the wide format to the long format; RColorBrewer for defining color palettes; directlabels for label positioning; classInt for computing class intervals; plotly, googleVis, and gridSVG for interactive graphics.

5.1 Polylines

Our first approach is to display the entire data in a panel with a scatterplot using country names as the grouping factor. Points of each country are connected with polylines to reveal the time evolution (Figure 5.1).

```
library(zoo)

load('data/CO2.RData')

## lattice version
xyplot(GNI.capita ~ CO2.capita, data = CO2data,
       xlab = "Carbon dioxide emissions (metric tons per capita)",
       ylab = "GNI per capita, PPP (current international $)",
       groups = Country.Name, type = 'b')

## ggplot2 version
ggplot(data = CO2data, aes(x = CO2.capita, y = GNI.capita,
                    color = Country.Name)) +
   xlab("Carbon dioxide emissions (metric tons per capita)") +
   ylab("GNI per capita, PPP (current international $)") +
   geom_point() + geom_path() + theme_bw()
```

Three improvements can be added to this graphical result:

1. Define a better palette to enhance visual discrimination between countries.

2. Display time information with labels to show year values.

3. Label each polyline with the country name instead of a legend.

5.1.1 Choosing Colors

The Country.Name categorical variable will be encoded with a qualitative palette, namely the first five colors of Set1 palette[1] from the RColorBrewer package (Neuwirth 2014). Because there are more countries than colors, we have to repeat some colors to complete the number of levels of the variable Country.Name. The result is a palette with non-unique colors, and thus some countries will share the same color. This is not a problem because the curves will be labeled, and countries with the same color will be displayed at enough distance.

---

[1]http://colorbrewer2.org/

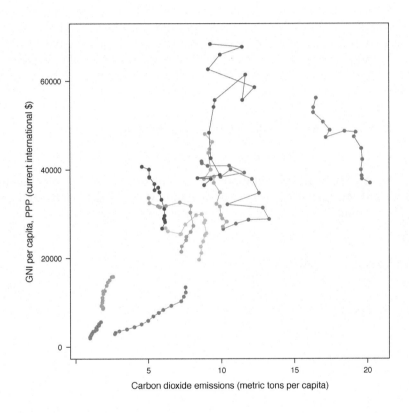

FIGURE 5.1: GNI per capita versus CO_2 emissions per capita (lattice version).

```
library(RColorBrewer)

nCountries <- nlevels(CO2data$Country.Name)
pal <- brewer.pal(n = 5, 'Set1')
pal <- rep(pal, length = nCountries)
```

Adjacent colors of this palette are chosen to be easily distinguishable. Therefore, the connection between colors and countries must be in such a way that nearby lines are encoded with adjacent colors of the palette.

A simple approach is to calculate the annual average of the variable to be represented along the x-axis (CO2.capita), and extract colors from the palette according to the order of this value.

```
## Rank of average values of CO2 per capita
CO2mean <- aggregate(CO2.capita ~ Country.Name,
                     data = CO2data, FUN = mean)
palOrdered <- pal[rank(CO2mean$CO2.capita)]
```

A more sophisticated solution is to use the ordered results of a hierarchical clustering of the time evolution of the CO_2 per capita values (Figure 5.2). The data is extracted from the original CO_2 data.frame.

```
library(reshape2)

CO2capita <- CO2data[, c('Country.Name',
                         'Year',
                         'CO2.capita')]
CO2capita <- dcast(CO2capita, Country.Name ~ Year)

summary(CO2capita)

Using CO2.capita as value column: use value.var to override.
  Country.Name 2000 2001 2002
 Brazil :1 Min. : 0.9799 Min. : 0.9717 Min. : 0.9674
 China  :1 1st Qu.: 3.5093 1st Qu.: 3.5949 1st Qu.: 3.7725
 Finland:1 Median : 7.8681 Median : 7.9634 Median : 7.9849
 France :1 Mean  : 7.6468 Mean  : 7.7977 Mean  : 7.7982
 Germany:1 3rd Qu.: 9.7802 3rd Qu.:10.0960 3rd Qu.: 9.6920
 Greece :1 Max.  :20.1788 Max.  :19.6365 Max.  :19.6134
 (Other):4
       2003 2004 2005 2006
 Min. : 0.9924 Min. : 1.025 Min. : 1.069 Min. : 1.122
 1st Qu.: 4.1721 1st Qu.: 4.559 1st Qu.: 4.917 1st Qu.: 5.212
 Median : 8.1860 Median : 8.388 Median : 8.539 Median : 8.356
```

```
Mean : 8.1468 Mean : 8.146 Mean : 7.948 Mean : 8.163
3rd Qu.: 9.9536 3rd Qu.: 9.746 3rd Qu.: 9.545 3rd Qu.: 9.808
Max. :19.5641 Max. :19.658 Max. :19.592 Max. :19.094


     2007 2008 2009 2010
Min.  : 1.193 Min.  : 1.310 Min.  : 1.432 Min.  : 1.397
1st Qu.: 5.443 1st Qu.: 5.693 1st Qu.: 5.581 1st Qu.: 5.526
Median : 8.407 Median : 7.914 Median : 7.247 Median : 7.050
Mean  : 8.139 Mean  : 8.082 Mean  : 7.665 Mean  : 7.947
3rd Qu.: 9.553 3rd Qu.:10.354 3rd Qu.: 9.671 3rd Qu.:11.001
Max. :19.218 Max. :18.462 Max. :17.158 Max. :17.442


     2011 2012 2013 2014
Min.  : 1.477 Min.  : 1.598 Min.  : 1.591 Min.  : 1.730
1st Qu.: 5.255 1st Qu.: 5.222 1st Qu.: 5.068 1st Qu.: 4.688
Median : 7.216 Median : 7.336 Median : 6.947 Median : 6.862
Mean  : 7.475 Mean  : 7.387 Mean  : 7.396 Mean  : 7.097
3rd Qu.: 9.125 3rd Qu.: 9.168 3rd Qu.: 9.213 3rd Qu.: 8.832
Max. :16.972 Max. :16.304 Max. :16.316 Max. :16.494
```

```
hCO2 <- hclust(dist(CO2capita[, -1]))

oldpar <- par(mar = c(0, 2, 0, 0) + .1)
plot(hCO2, labels = CO2capita$Country.Name,
    xlab = '', ylab = '', sub = '', main = '')
par(oldpar)
```

The colors of the palette are assigned to each country with match, which returns a vector of the positions of the matches of the country names in alphabetical order in the country names ordered according to the hierarchical clustering.

```
idx <- match(levels(CO2data$Country.Name),
        CO2capita$Country.Name[hCO2$order])
palOrdered <- pal[idx]
```

It must be highlighted that this palette links colors with the levels of Country.Name (country names in alphabetical order), which is exactly what the groups argument provides. The following code produces a curve for each country using different colors to distinguish them.

```
## simpleTheme encapsulates the palette in a new theme for xyplot
myTheme <- simpleTheme(pch = 19, cex = 0.6, col = palOrdered)
```

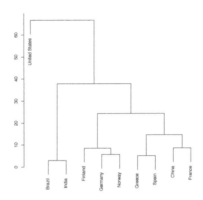

FIGURE 5.2: Hierarchical clustering of the time evolution of CO_2 per capita values.

```
## lattice version
pCO2.capita <- xyplot(GNI.capita ~ CO2.capita,
                 data = CO2data,
                 xlab = "Carbon dioxide emissions (metric tons
                     per capita)",
                 ylab = "GNI per capita, PPP (current
                     international $)",
                 groups = Country.Name,
                 par.settings = myTheme,
                 type = 'b')

## ggplot2 version
gCO2.capita <- ggplot(data = CO2data,
                 aes(x = CO2.capita,
                    y = GNI.capita,
                    color = Country.Name)) +
    geom_point() + geom_path() +
    scale_color_manual(values = palOrdered, guide = FALSE) +
    xlab('CO2 emissions (metric tons per capita)') +
    ylab('GNI per capita, PPP (current international $)') +
    theme_bw()
```

5.1.2   Labels to Show Time Information

This result can be improved with labels displaying the years to show the time evolution. A panel function with panel.text to print the year labels and panel.superpose to display the lines for each group is a solution. In the panel function, subscripts is a vector with the integer indices representing the rows of the data.frame to be displayed in the panel.

```
xyplot(GNI.capita ~ CO2.capita,
       data = CO2data,
       xlab = "Carbon dioxide emissions (metric tons per capita)",
       ylab = "GNI per capita, PPP (current international $)",
       groups = Country.Name,
       par.settings = myTheme,
       type = 'b',
       panel = function(x, y, ..., subscripts, groups){
          panel.text(x, y, ...,
                     labels = CO2data$Year[subscripts],
                     pos = 2, cex = 0.5, col = 'gray')
          panel.superpose(x, y, subscripts, groups,...)
       })
```

The same result with a clearer code is obtained with the combination of +.trellis, glayer_ and panel.text. Using glayer_ instead of glayer, we ensure that the labels are printed below the lines.

```
## lattice version
pCO2.capita <- pCO2.capita +
   glayer_(panel.text(...,
                      labels = CO2data$Year[subscripts],
                      pos = 2, cex = 0.5, col = 'gray'))

## ggplot2 version
gCO2.capita <- gCO2.capita + geom_text(aes(label = Year),
                                        colour = 'gray',
                                        size = 2.5,
                                        hjust = 0, vjust = 0)
```

5.1.3   Country Names: Positioning Labels

The common solution to link each curve with the group value is to add a legend. However, a legend can be confusing with too many items. In addition, the reader must carry out a complex task: Choose the line, memorize its color, search for it in the legend, and read the country name.

A better approach is to label each line using nearby text with the same color encoding. A suitable method is to place the labels avoiding the overlapping between labels and lines. The package directlabels (Hocking 2017) includes a wide repertory of positioning methods to cope with overlapping. The main function, direct.label, is able to determine a suitable method for each plot, although the user can choose a different method from the collection or even define a custom method. For the pCO2.capita object, the best results are obtained with extreme.grid (Figure 5.3).

```
library(directlabels)

## lattice version
direct.label(pCO2.capita,
            method = 'extreme.grid')

## ggplot2 version
direct.label(gCO2.capita, method = 'extreme.grid')
```

5.2 A Panel for Each Year

Time can be used as a conditioning variable (as shown in previous sections) to display subsets of the data in different panels. Figure 5.4 is produced with the same code as in Figure 5.1, now including |factor(Year) in the lattice version and facet_wrap(~ Year) in the ggplot2 version.

```
## lattice version
xyplot(GNI.capita ~ CO2.capita | factor(Year),
       data = CO2data,
       xlab = "Carbon dioxide emissions (metric tons per capita)",
       ylab = "GNI per capita, PPP (current international $)",
       groups = Country.Name, type = 'b',
       auto.key = list(space = 'right'))

## ggplot2 version
ggplot(data = CO2data,
       aes(x = CO2.capita,
           y = GNI.capita,
           colour = Country.Name)) +
    facet_wrap(~ Year) + geom_point(pch = 19) +
    xlab('CO2 emissions (metric tons per capita)') +
    ylab('GNI per capita, PPP (current international $)') +
    theme_bw()
```

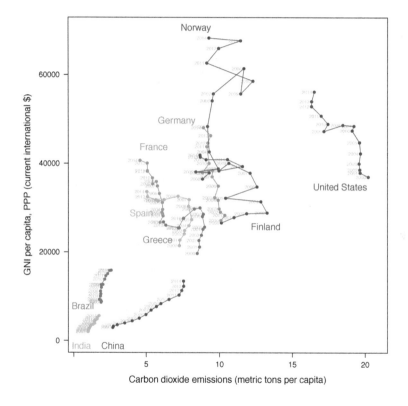

FIGURE 5.3: CO$_2$ emissions versus GNI per capita. Labels are placed with the extreme.grid method of the directlabels package.

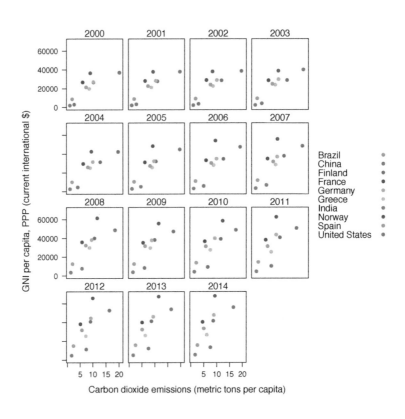

FIGURE 5.4: CO$_2$ emissions versus GNI per capita with a panel for each year.

Because the grouping variable, Country.Name, has many levels, the legend is not very useful. Once again, point labeling is recommended (Figure 5.5).

```
## lattice version
xyplot(GNI.capita ~ CO2.capita | factor(Year),
       data = CO2data,
       xlab = "Carbon dioxide emissions (metric tons per capita)",
       ylab = "GNI per capita, PPP (current international $)",
       groups = Country.Name, type = 'b',
       par.settings = myTheme) +
    glayer(panel.pointLabel(x, y,
                      labels = group.value,
                      col = palOrdered[group.number],
                      cex = 0.7))
```

5.2.1 ✎Using Variable Size to Encode an Additional Variable

Instead of using simple points, we can display circles of different radius to encode a new variable. This new variable is CO2.PPP, the ratio of CO_2 emissions to the Gross Domestic Product with purchasing power parity (PPP) estimations.

To use this numeric variable as an additional grouping factor, its range must be divided into different classes. The typical solution is to use cut to coerce the numeric variable into a factor whose levels correspond to uniform intervals, which could be unrelated to the data distribution. The classInt package (Bivand 2017) provides several methods to partition data into classes based on natural groups in the data distribution.

```
library(classInt)
z <- CO2data$CO2.PPP
intervals <- classIntervals(z, n = 4, style = 'fisher')
```

Although the functions of this package are mainly intended to create color palettes for maps, the results can also be associated to point sizes. cex.key defines the sequence of sizes (to be displayed in the legend) associated with each CO2.PPP using the findCols function.

```
nInt <- length(intervals$brks) - 1
cex.key <- seq(0.5, 1.8, length = nInt)

idx <- findCols(intervals)
CO2data$cexPoints <- cex.key[idx]
```

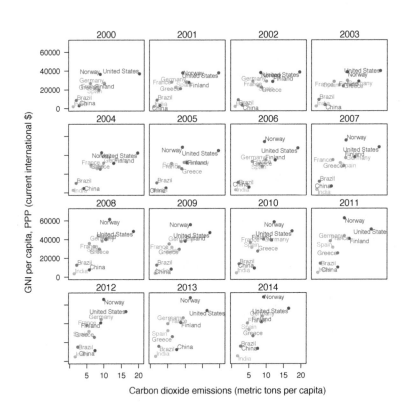

FIGURE 5.5: CO$_2$ emissions versus GNI per capita with a panel for each year.

The graphic will display information on two variables (GNI.capita and CO2.capita in the vertical and horizontal axes, respectively) with a conditioning variable (Year) and two grouping variables (Country.Name, and CO2.PPP through cexPoints) (Figure 5.6).

```
ggplot(data = CO2data,
       aes(x = CO2.capita,
           y = GNI.capita,
           colour = Country.Name)) +
    facet_wrap(~ Year) +
    geom_point(aes(size = cexPoints), pch = 19) +
    xlab('Carbon dioxide emissions (metric tons per capita)') +
    ylab('GNI per capita, PPP (current international $)') +
    theme_bw()
```

The auto.key mechanism of the lattice version is not able to cope with two grouping variables. Therefore, the legend, whose main components are the labels (intervals) and the point sizes (cex.key), should be defined manually (Figure 5.7).

```
op <- options(digits = 2)
tab <- print(intervals)
options(op)

key <- list(space = 'right',
            title = expression(CO[2]/GNI.PPP),
            cex.title = 1,
            ## Labels of the key are the intervals strings
            text = list(labels = names(tab), cex = 0.85),
            ## Points sizes are defined with cex.key
            points = list(col = 'black',
                          pch = 19,
                          cex = cex.key,
                          alpha = 0.7))

xyplot(GNI.capita ~ CO2.capita|factor(Year), data = CO2data,
       xlab = "Carbon dioxide emissions (metric tons per capita)",
       ylab = "GNI per capita, PPP (current international $)",
       groups = Country.Name, key = key, alpha = 0.7,
       panel = panel.superpose,
       panel.groups = function(x, y,
           subscripts, group.number, group.value, ...){
           panel.xyplot(x, y,
                        col = palOrdered[group.number],
```

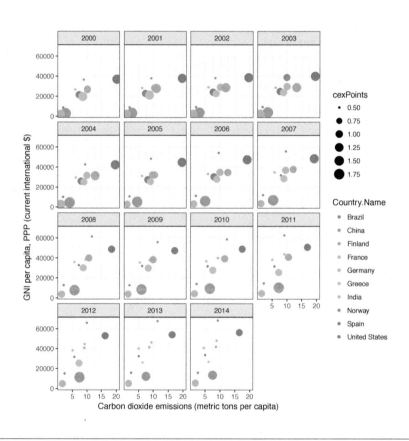

FIGURE 5.6: CO_2 emissions versus GNI per capita for different intervals of the ratio of CO_2 emissions to the GDP PPP estimations.

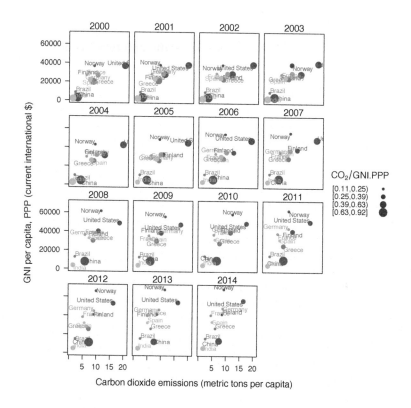

FIGURE 5.7: CO_2 emissions versus GNI per capita for different intervals of the ratio of CO_2 emissions to the GDP PPP estimations.

```
            cex = CO2data$cexPoints[subscripts])
panel.pointLabel(x, y, labels = group.value,
            col = palOrdered[group.number],
            cex = 0.7)
    }
)
```

5.3 Interactive Graphics: Animation

Previous sections have been focused on static graphics. This section describes several solutions to display the data through animation with interactive functionalities.

Gapminder [2] is an independent foundation based in Stockholm, Sweden. Its mission is "to debunk devastating myths about the world by offering free access to a fact-based world view." They provide free online tools, data, and videos "to better understand the changing world." The initial development of Gapminder was the Trendalyzer software, used by Hans Rosling in several sequences of his documentary "The Joy of Stats."

The information visualization technique used by Trendalyzer is an interactive bubble chart. By default it shows five variables: two numeric variables on the vertical and horizontal axes, bubble size and color, and a time variable that may be manipulated with a slider. The software uses brushing and linking techniques for displaying the numeric value of a highlighted country.

This software was acquired by Google in 2007, and is now available as a Motion Chart gadget and as the Public Data Explorer.

We will mimic the Trendalyzer/Motion Chart solution, using traveling bubbles of different colors and with radius proportional to the values of the variable $CO2.PPP$. Previously, you should watch the magistral video "200 Countries, 200 Years, 4 Minutes"[3].

Three packages are used here: googleVis, plotly, and gridSVG.

5.3.1 plotly

The plotly package has already been used in Section 3.4.3 to create an interactive graphic representing time in the x-axis. In this section this package produces an animation piping the result of the plot_ly and add_markers functions through the animation_slider function.

Variables CO2.capita and GNI.capita are represented in the x-axis and y-axis, respectively.

```
library(plotly)
p <- plot_ly(CO2data,
             x = ~CO2.capita,
             y = ~GNI.capita)
```

---

[2]http://www.gapminder.org/

[3]http://www.gapminder.org/videos/200-years-that-changed-the-world-bbc/

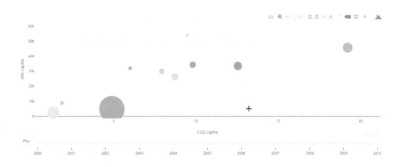

FIGURE 5.8: Snapshot of a Motion Chart produced with plotly.

CO2.PPP is encoded with the circle sizes, while Country.Name is represented with colours and labels.

```
p <- add_markers(p,
                size = ~CO2.PPP,
                color = ~Country.Name,
                text = ~Country.Name, hoverinfo = "text",
                ids = ~Country.Name,
                frame = ~Year,
                showlegend = FALSE)
```

Finally, animation is created with animation_opts, to customize the frame and transition times, and with animation_slider to define the slider. Figure 5.8 is an snapshot of this animation.

```
p <- animation_opts(p,
                frame = 1000,
                transition = 800,
                redraw = FALSE)

p <- animation_slider(p,
                currentvalue = list(prefix = "Year "))

p
```

5.3.2 googleVis

The googleVis package (Gesmann and Castillo 2011) is an interface between R and the Google Visualisation API. With its gvisMotionChart func-

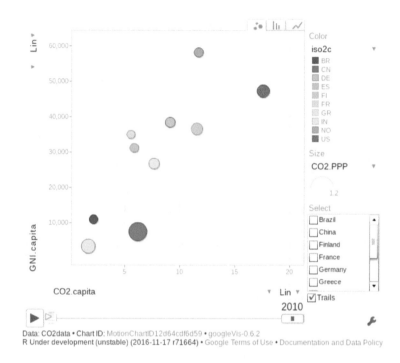

FIGURE 5.9: Snapshot of a Motion Chart produced with googleVis.

tion it is easy to produce a Motion Chart that can be displayed using a browser with Flash enabled (Figure 5.9).

```
library(googleVis)

pgvis <- gvisMotionChart(CO2data,
                xvar = 'CO2.capita',
                yvar = 'GNI.capita',
                sizevar = 'CO2.PPP',
                idvar = 'Country.Name',
                timevar = 'Year')
```

Although the gvisMotionChart is quite easy to use, the global appearance and behavior are completely determined by Google API[4]. Moreover,

---

[4]You should read the Google API Terms of Service before using googleVis: https://developers.google.com/terms/

you should carefully read their Terms of Use before using it for public distribution.

5.3.3 ✆gridSVG

The final solution to create an animation is based on the function grid.animate of the gridSVG package.

The first step is to draw the initial state of the bubbles. Their colors are again defined by the palOrdered palette (line 15), although the adjust-color function is used for a ligther fill color. Because there will not be a legend, there is no need to define class intervals, and thus the radius is directly proportional to the value of CO2data$CO2.PPP (line 16).

```
1  library(gridSVG)
2  library(grid)
3
4  xyplot(GNI.capita ~ CO2.capita,
5         data = CO2data,
6         xlab = "Carbon dioxide emissions (metric tons per capita)",
7         ylab = "GNI per capita, PPP (current international $)",
8         subset = Year==2000,
9         groups = Country.Name,
10        ## The limits of the graphic are defined
11        ## with the entire dataset
12        xlim = extendrange(CO2data$CO2.capita),
13        ylim = extendrange(CO2data$GNI.capita),
14        panel = function(x, y, ..., subscripts, groups) {
15            color <- palOrdered[groups[subscripts]]
16            radius <- CO2data$CO2.PPP[subscripts]
17            ## Size of labels
18            cex <- 1.1*sqrt(radius)
19            ## Bubbles
20            grid.circle(x, y, default.units = "native",
21                        r = radius*unit(.25, "inch"),
22                        name = trellis.grobname("points", type = "panel
                            "),
23                        gp = gpar(col = color,
24                            ## Fill color ligther than border
25                            fill = adjustcolor(color, alpha = .5),
26                            lwd = 2))
27            ## Country labels
28            grid.text(label = groups[subscripts],
29                      x = unit(x, 'native'),
```

83

```
30                      ## Labels above each bubble
31                      y = unit(y, 'native') + 1.5 * radius *unit(.25,
                             'inch'),
32                      name = trellis.grobname('labels', type = 'panel'
                             ),
33                      gp = gpar(col = color, cex = cex))
34              })
```

From this initial state, grid.animate creates a collection of animated
graphical objects with the result of animUnit (lines 9, 11, 14 and 18). This
function produces a set of values that will be interpreted by grid.animate
as intermediate states of a feature of the graphical object (lines 21 and 29).
Thus, the bubbles will travel across the values defined by x_points and
y_points, while their labels will use x_points and x_labels.

The use of rep=TRUE ensures that the animation will be repeated indefinitely (lines 27 and 34).

```
1   ## Duration in seconds of the animation
2   duration <- 20
3
4   nCountries <- nlevels(CO2data$Country.Name)
5   years <- unique(CO2data$Year)
6   nYears <- length(years)
7
8   ## Intermediate positions of the bubbles
9   x_points <- animUnit(unit(CO2data$CO2.capita, 'native'),
10                   id = rep(seq_len(nCountries), each = nYears))
11  y_points <- animUnit(unit(CO2data$GNI.capita, 'native'),
12                   id = rep(seq_len(nCountries), each = nYears))
13  ## Intermediate positions of the labels
14  y_labels <- animUnit(unit(CO2data$GNI.capita, 'native') +
15                   1.5 * CO2data$CO2.PPP * unit(.25, 'inch'),
16                   id = rep(seq_len(nCountries), each = nYears))
17  ## Intermediate sizes of the bubbles
18  size <- animUnit(CO2data$CO2.PPP * unit(.25, 'inch'),
19              id = rep(seq_len(nCountries), each = nYears))
20
21  grid.animate(trellis.grobname("points", type = "panel",
22                      row = 1, col = 1),
23          duration = duration,
24          x = x_points,
25          y = y_points,
26          r = size,
27          rep = TRUE)
```

```
28
29   grid.animate(trellis.grobname("labels", type = "panel",
30                        row = 1, col = 1),
31            duration = duration,
32            x = x_points,
33            y = y_labels,
34            rep = TRUE)
```

A bit of interactivity can be added with the grid.hyperlink function. For example, the following code adds the corresponding Wikipedia link to a mouse click on each bubble.

```
countries <- unique(CO2data$Country.Name)
URL <- paste('http://en.wikipedia.org/wiki/', countries, sep = ''
    )
grid.hyperlink(trellis.grobname('points', type = 'panel', row =
    1, col = 1),
              URL, group = FALSE)
```

Finally, the time information: The year is printed in the lower right corner, using the visibility attribute of an animated textGrob object to show and hide the values.

```
visibility <- matrix("hidden", nrow = nYears, ncol = nYears)
diag(visibility) <- "visible"
yearText <- animateGrob(garnishGrob(textGrob(years, .9, .15,
                                    name = "year",
                                    gp = gpar(cex = 2, col = "
                                        grey")),
                            visibility = "hidden"),
                    duration = 20,
                    visibility = visibility,
                    rep = TRUE)
grid.draw(yearText)
```

The SVG file produced with grid.export is available at the website of the book (Figure 5.10). Because this animation does not trace the paths, Figure 5.3 provides this information as a static complement.

```
grid.export("figs/bubbles.svg")
```

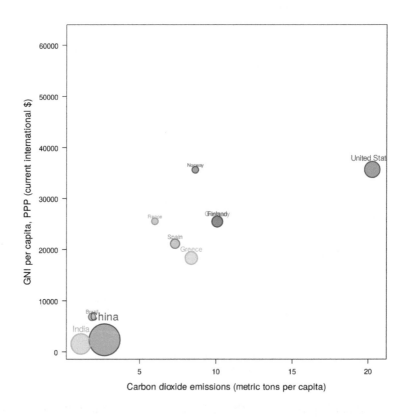

FIGURE 5.10: Animated bubbles produced with gridSVG.

Chapter 6

About the Data

6.1 SIAR

The Agroclimatic Information System for Irrigation (SIAR) (MARM 2011) is a free-download database operating since 1999, covering the majority of the irrigated area of Spain. This network belongs to the Ministry of Agriculture, Food and Environment of Spain, as a tool to predict and study meteorological variables for agriculture. SIAR is composed of twelve regional centers and a national center, aiming to centralize and depurate measurements from the stations of the network. Figure 6.1 displays the stations over an altitude map. Some stations from the complete network have been omitted, due to difficulties accessing their coordinates or to incomplete or spurious data series[1].

6.1.1 Daily Data of Different Meteorological Variables

As an example of multiple time series with different scales, we will use eight years (from January 2004 to December 2011) of daily data corresponding to several meteorological variables measured at the SIAR station located at Aranjuez (Madrid, Spain) available on the SIAR webpage[2].

---

[1]The name and location data of these stations are available at the GitHub repository of the paper (Antonanzas-Torres, Cañizares, and Perpiñán 2013).

[2]http://eportal.magrama.gob.es/websiar

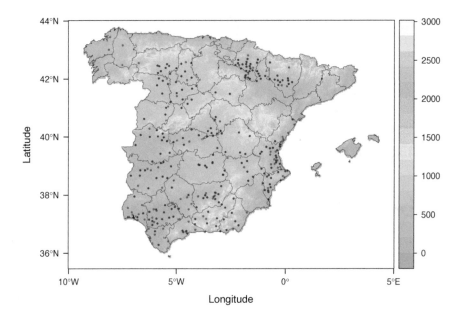

FIGURE 6.1: Meteorological stations of the SIAR network. The color key indicates the altitude (meters).

The aranjuez.gz file, available in the data folder of the book repository, contains this information with several meteorological variables: average, maximum, and minimum ambient temperature; average and maximum humidity; average and maximum wind speed; rainfall; solar radiation on the horizontal plane; and evotranspiration.

The read.zoo from the zoo package accepts this string and downloads the data to construct a zoo object. Several arguments are passed directly to read.table (header, skip, etc.) and are detailed conveniently on the help page of this function. The index.column is the number of the column with the time index, and format defines the date format of this index.

```
library(zoo)

aranjuez <- read.zoo("data/aranjuez.gz",
                index.column = 3, format = "%d/%m/%Y",
                fileEncoding = 'UTF-16LE',
                header = TRUE, fill = TRUE,
                sep = ';', dec = ",", as.is = TRUE)
```

```
aranjuez <- aranjuez[, -c(1:4)]

names(aranjuez) <- c('TempAvg', 'TempMax', 'TempMin',
                     'HumidAvg', 'HumidMax',
                     'WindAvg', 'WindMax',
                     'Radiation', 'Rain', 'ET')
```

From the summary it is clear that parts of these time series include erroneous outliers that can be safely removed:

```
aranjuezClean <- within(as.data.frame(aranjuez),{
    TempMin[TempMin > 40] <- NA
    HumidMax[HumidMax > 100] <- NA
})

aranjuez <- zoo(aranjuezClean, index(aranjuez))
```

6.1.2 Solar Radiation Measurements from Different Locations

As an example of multiple time series with the same scale, we will use data of daily solar radiation measurements from different locations.

Daily solar radiation incident on the horizontal plane is registered by meterological stations and estimated from satellite images. This meteorological variable is important for a wide variety of scientific disciplines and engineering applications. Its variations and trends, dependent on the location (mainly latitude, and also longitude and altitude) and on time (day of the year), have been analyzed and modeled in a huge collection of papers and reports. In this section we will focus our attention on the time evolution of the solar radiation. The spatial distribution and the spatio-temporal behavior will be the subject of later sections.

The stations of the SIAR network include first-class pyranometers according to the World Meteorological Organization (WMO), whose absolute accuracy is within $\pm 5\%$ and is typically lower than $\pm 3\%$. Solar irradiance is recorded every 15 minutes and then collated through a datalogger within the station to generate the daily irradiation, which is later sent to the regional and national centers.

The file navarra.RData contains daily solar radiation data of 2011 from the meteorological stations of Navarra, Spain. The names of the dataset are the abbreviations of each station name.

```
library(zoo)

load('data/navarra.RData')
```

6.2   Unemployment in the United States

As an example of time series that can be displayed both in individual and in aggregate, we will use the unemployment data in the United States. The information on unemployed persons by industry and class of worker is available in Table A-14 published by the Bureau of Labor Statistics[3].

The dataset arranges the information with a row for each category (Series.ID) and a column for each monthly value. In addition, there are columns with the annual summaries (annualCols). We rearrange this data.frame, dropping the Series.ID and the annual columns, and transpose the data.

```
unemployUSA <- read.csv('data/unemployUSA.csv')
nms <- unemployUSA$Series.ID
##columns of annual summaries
annualCols <- 14 + 13*(0:12)
## Transpose. Remove annual summaries
unemployUSA <- as.data.frame(t(unemployUSA[,-c(1, annualCols)]))
## First 7 characters can be suppressed
names(unemployUSA) <- substring(nms, 7)
```

With the transpose, the column names of the original data set are now the row names of the data.frame. The as.yearmon function of the zoo package converts the character vector of names into a yearmon vector, a class for representing monthly data. With Sys.setlocale("LC_TIME", 'C') we ensure that month abbreviations (%b) are correctly interpreted in a non-English locale. This vector is the time index of a new zoo object.

```
library(zoo)

Sys.setlocale("LC_TIME", 'C')
idx <- as.yearmon(row.names(unemployUSA), format = '%b.%Y')
unemployUSA <- zoo(unemployUSA, idx)
```

Finally, those rows with NA values are removed.

```
unemployUSA <- unemployUSA[complete.cases(unemployUSA), ]
```

---

[3]http://www.bls.gov/webapps/legacy/cpsatab14.htm

6.3   Gross National Income and CO_2 Emissions

The catalog data of the World Bank Open Data initiative includes the World Development Indicators (WDI)[4]. Among them we will analyze the evolution of the relationship between Gross National Income (GNI) and CO_2 emissions for a set of countries from 2000 to 2014[5]. The package WDI is able to search and download these data series.

```
library(WDI)

CO2data <- WDI(indicator = c('EN.ATM.CO2E.PC', 'EN.ATM.CO2E.PP.GD
        ',

                    'NY.GNP.MKTP.PP.CD', 'NY.GNP.PCAP.PP.CD'),
            start = 2000, end = 2014,
            country = c('BR', 'CN', 'DE',
                    'ES', 'FI', 'FR',
                    'GR', 'IN', 'NO',
                    'US'))

names(CO2data) <- c('iso2c', 'Country.Name', 'Year',
                    'CO2.capita', 'CO2.PPP',
                    'GNI.PPP', 'GNI.capita')
```

Only two minor modifications are needed: Remove rows with missing values (using complete.cases), and convert the Country.Name column into a factor. This first modification will save problems when displaying the time series, and the factor conversion will be useful for grouping.

```
CO2data <- CO2data[complete.cases(CO2data), ]

CO2data$Country.Name <- factor(CO2data$Country.Name)
```

---

[4]http://databank.worldbank.org/data/reports.aspx?source=world-development-indicators

[5]The analysis is restricted until 2014 because the database has several gaps in the range from 2015 to 2017.

Part II

Spatial Data

Chapter 7

Displaying Spatial Data: Introduction

Spatial data (also known as geospatial data) are directly or indirectly referenced to a location on the surface of the Earth. Their spatial reference is composed of coordinate values and a system of reference for these coordinates. Spatial data are often accessed, manipulated, or analyzed through Geographic Information Systems (GIS).

Real objects represented by GIS data can be divided into two abstractions: discrete objects (e.g., a road or a river) represented with vector data (points, lines, and polygons), and continuous fields (such as elevation or solar radiation) represented with raster data. The sp and sf packages are the preferred option to use vector data in R, and the raster package is the choice for raster data [1].

This part exposes several examples where vector and raster data are displayed. These examples make use of several datasets (available at the book website) described in Chapter 13.

On the one hand, the Chapters 8, 9, and 10 focus on thematic maps, that display a specific variable commonly using geographic data such as coastlines, boundaries, and places as points of reference for the variable

---

[1] Although sp, sf, and raster are the most important packages, there are an increasing number of packages designed to work with spatial data. They are summarized in the corresponding CRAN Task View. Read Section 7.2 for details.

being mapped. These maps provide specific information about particular locations or areas (proportional symbol mapping and choropleth maps) and information about spatial patterns (isarithmic and raster maps).

On the other hand, the Chapter 12 focuses on reference maps, to show geographic location of features, and on physical maps, to show the landscape and features of a place.

7.1 Packages

The CRAN Tasks View "Analysis of Spatial Data" [2] summarizes the packages for reading, vizualizing, and analyzing spatial data. This section provides a brief introduction to sp, sf, raster, rasterVis, maptools, rgdal, gstat, and maps. Most of the information has been extracted from their vignettes, webpages, and help pages. You should read them for detailed information.

7.1.1 sp

The sp package (E. J. Pebesma and Bivand 2005) provides classes and methods for dealing with spatial data in R. The spatial data classes implemented are points (SpatialPoints), grids (SpatialPixels and Spatial-Grid), lines (Line, Lines and SpatialLines), rings, and polygons (Polygon, Polygons, and SpatialPolygons), each of them without data or with data (for example, SpatialPointsDataFrame or SpatialLinesDataFrame)[3].

Selecting, retrieving, or replacing certain attributes in spatial objects with data is done using standard methods:

- [selects rows (items) and columns in the data.frame.

- [[selects a column from the data.frame

- [[<- assigns or replaces values to a column in the data.frame.

A number of spatial methods are available for the classes in sp:

---

[2]http://CRAN.R-project.org/view=Spatial

[3]The asterisk is commonly used as a wildcard character to denote subsets of classes. Thus, SpatialLines* comprises SpatialLines and SpatialLinesDataFrame classes. Moreover, Spatial* represents all the classes defined by the sp package.

- `coordinates(object) <- value` sets spatial coordinates to create spatial data. It promotes a `data.frame` into a `SpatialPointsData-Frame`. *value* may be specified by a formula, a character vector, or a numeric matrix or `data.frame` with the actual coordinates.

- `coordinates(object, ...)` returns a matrix with the spatial coordinates. If used with `SpatialPolygons` it returns a matrix with the centroids of the polygons.

- `bbox` returns a matrix with the coordinates bounding box.

- `proj4string(object)` and `proj4string(object) <- value` retrieve or set projection attributes on spatial classes.

- `spTransform` transforms from one coordinate reference system (geographic projection) to another (requires package `rgdal`).

- `spplot` plots attributes combined with spatial data: Points, lines, grids, polygons.

7.1.2 sf

The `sf` package (E. Pebesma 2018), the long term successor of `sp`, implements simple features in R. Simple features is an open (OGC and ISO) interface standard for access and manipulation of spatial vector data (points, lines, polygons).

This package represents simple features using simple data structures, commonly `data.frame` objects. Feature geometries are stored in a data.frame column, using a list-column because geometries are not single-valued. The length of this list is equal to the number of records in the `data.frame`, with the simple feature geometry of that feature in each element of the list.

`sf` implements three classes to represent simple features:

- `sf`, a `data.frame` with feature attributes and feature geometries. It contains

- `sfc`,the list-column with the geometries for each feature (record), which is composed of

- `sfg`, the feature geometry of an individual simple feature.

All functions and methods in `sf` that operate on spatial data are prefixed by `st_` (spatial and temporal). For the purposes of this book, the most important are:

- `st_read`, `st_write`, for reading and writing spatial data, respectively.

- `st_transform` for coordinate reference system transformations.

- `st_as_sf.*`, a family of conversions functions between sp and sf.

The sf package implements `plot` methods for displaying data using base graphics. Besides, this package provides a number of methods for conversion to grob objects in order to display these objects with packages working with the `grid` system (`lattice` and `ggplot2`). Finally, the ggplot2 version[4] to be released after 2.2.1 (on CRAN at the time of writing this book) contains the `geom_sf` geom, designed for sf objects.

7.1.3 raster

The `raster` package (R. J. Hijmans 2017) has functions for creating, reading, manipulating, and writing raster data. The package provides general raster data manipulation functions. The package also implements raster algebra and most functions for raster data manipulation that are common in Geographic Information Systems (GIS).

The raster package can work with raster datasets stored on disk if they are too large to be loaded into memory. The package can work with large files because the objects it creates from these files only contain information about the structure of the data, such as the number of rows and columns, the spatial extent, and the filename, but it does not attempt to read all the cell values in memory. In computations with these objects, the data are processed in chunks.

The package defines a number of S4 classes. `RasterLayer, RasterBrick`, and `RasterStack` are the most important:

- A `RasterLayer` object represents single-layer (variable) raster data. It can be created with the function `raster`. This function is able to create a `RasterLayer` from another object, including another `Raster*` object[5], or from a `SpatialPixels*` and `SpatialGrid*` object, or even a matrix. In addition, it can create a `RasterLayer` reading data from a file. The raster package can use raster files in several formats, some of them via the `rgdal` package. Supported formats for reading include GeoTIFF, ESRI, ENVI, and ERDAS.

---

[4]The development version can be installed with the remotes package: `remotes::install_github("tidyverse/ggplot2")`.

[5]The notation `Raster*` represents all the classes of Raster objects: `RasterLayer`, `RasterStack`, and `RasterBrick`.

- RasterBrick and RasterStack are classes for multilayer data. A RasterStack is a list of RasterLayer objects with the same spatial extent and resolution. A RasterStack can be formed with a collection of files in different locations or even mixed with RasterLayer objects that only exist in memory. A RasterBrick is truly a multilayered object, and processing it can be more efficient than processing a RasterStack representing the same data.

The raster package defines a number of methods for raster algebra with Raster* objects: arithmetic operators, logical operators, and functions such as abs, round, ceiling, floor, trunc, sqrt, log, log10, exp, cos, sin, max, min, range, prod, sum, any, and all. In these functions, Raster* objects can be mixed with numbers.

There are several functions to modify the content or the spatial extent of Raster* objects, or to combine Raster* objects:

- The crop function takes a geographic subset of a larger Raster* object. trim crops a RasterLayer by removing the outer rows and columns that only contain NA values. extend adds new rows and/or columns with NA values.

- The merge function merges two or more Raster* objects into a single new object.

- projectRaster transforms values of a Raster* object to a new object with a different coordinate reference system.

- With overlay, multiple Raster* objects can be combined (for example, multiply them).

- mask removes all values from one layer that are NA in another layer, and cover combines two layers by taking the values of the first layer except where these are NA.

- calc computes a function for a Raster* object. With RasterLayer objects, another RasterLayer is returned. With multilayer objects the result depends on the function: With a summary function (sum, max, etc.), calc returns a RasterLayer object, and a RasterBrick object otherwise.

- stackApply computes summary layers for subsets of a RasterStack or RasterBrick.

- `cut` and `reclassify` replace ranges of values with single values.

- `zonal` computes zonal statistics, that is, summarizes a `Raster*` object using zones (areas with the same integer number) defined by another `RasterLayer`.

7.1.4 rasterVis

The `rasterVis` package (Perpiñán and R. Hijmans 2017) complements the `raster` package, providing a set of methods for enhanced visualization and interaction. This package defines visualization methods (`levelplot`) for quantitative data and categorical data, both for univariate and multivariate rasters.

It also includes several methods in the frame of the Exploratory Data Analysis approach: scatterplots with `xyplot`, histograms and density plots with `histogram` and `densityplot`, violin and boxplots with `bwplot`, and a matrix of scatterplots with `splom`.

On the other hand, this package is able to display vector fields using arrows, `vectorplot`, or with streamlines (Wegenkittl and Gröller 1997), `streamplot`. In this last method, for each point, *droplet*, of a jittered regular grid, a short streamline portion, *streamlet*, is calculated by integrating the underlying vector field at that point. The main color of each streamlet indicates local vector magnitude (slope). Streamlets are composed of points whose sizes, positions, and color degradation encode the local vector direction (aspect).

7.1.5 rgdal

The `rgdal` package (Bivand, Keitt, and Rowlingson 2017) provides bindings to the Geospatial Data Abstraction Library (GDAL) [6]. With `readOGR` and `readGDAL`, both GDAL raster and OGR vector map data can be imported into R, and GDAL raster data and OGR vector data can be exported with `writeGDAL` and `writeOGR`.

In addition, this package provides access to projection and transformation operations from the PROJ.4 library [7]. This package implements several `spTransform` methods providing transformation between datums and conversion between projections using PROJ.4 projection arguments.

---

[6] http://www.gdal.org/
[7] https://trac.osgeo.org/proj/

7.1.6 maptools

The maptools package (Bivand and Lewin-Koh 2017) provides a set of tools for manipulating geographic data. The package also provides interface wrappers for exchanging spatial objects with packages such as PB-Smapping, spatstat, maps, RArcInfo, Stata tmap, WinBUGS, Mondrian, and others. The main functions in the context of this book are:

- map2SpatialPolygons and map2SpatialLines may be used to convert map objects returned by the map function in the maps package to the classes defined in the sp package.

- spCbind provides cbind-like methods for Spatial*DataFrame and data.frame objects.

The topology operations on geometries performed by this package (for example, unionSpatialPolygons) use the package rgeos, an interface to the Geometry Engine Open Source (GEOS) [8].

7.1.7 gstat

The gstat package (E. J. Pebesma 2004) provides functions for geostatistical modeling, prediction, and simulation, including variogram modeling and simple, ordinary, universal, and external drift kriging.

Most of the functionality of this package is beyond the scope of this book. However, some functions must be mentioned:

- variogram calculates the sample variogram from data, or for the residuals if a linear model is given. vgm generates a variogram and fit.variogram fit ranges and/or sills from a variogram model to a sample variogram.

- krige is the function for simple, ordinary or universal kriging. gstat is the function for univariate or multivariate geostatistical prediction.

---

[8]http://trac.osgeo.org/geos/

7.1.8 maps

The maps (Becker, Wilks, Brownrigg, et al. 2017), mapdata (Becker, Wilks, and Brownrigg 2017), and mapproj (McIlroy et al. 2017) packages are useful to draw or create geographical maps. mapdata contains higher resolution databases, and mapproj converts latitude/longitude coordinates into projected coordinates.

7.2 Further Reading

- (Slocum 2005) and (Dent, Torguson, and Hodler 2008) are comprehensive books on thematic cartography and geovisualization. They include chapters devoted to data classification, scales, map projections, color theory, typography, and proportional symbol, choropleth, dasymetric, isarithmic, and multivariate mapping. Several resources are available at their accompanying websites [9].

- (Bivand, E. J. Pebesma, and Gomez-Rubio 2013) is the essential reference to work with spatial data in R. R. Bivand and E. Pebesma are the authors of the fundamental sp package, and they are the authors or maintainers of several important packages such as gstat, for geostatistical modeling, prediction, and simulation, rgdal, rgeos and maptools. Chapter 3 is devoted to the visualization of spatial data. Code, figures, and data of the book are available at the accompanying website [10].

- (Hengl 2009) is an open-access book with seven spatial data analysis exercises. The author is the creator and maintainer of the Spatial-Analyst webpage [11].

- The CRAN Tasks View "Analysis of Spatial Data" [12] summarizes the packages for reading, vizualizing, and analyzing spatial data. The packages in development published at R-Forge are listed in the "Spatial Data & Statistics" topic view [13]. The R-SIG-Geo mailing list [14] is a powerful resource for obtaining help.

---

[9]http://www.pearsonhighered.com/slocum3e/ and http://highered.mcgraw-hill.com/sites/0072943823/

[10]http://www.asdar-book.org/

[11]http://spatial-analyst.net

[12]http://CRAN.R-project.org/view=Spatial

[13]http://r-forge.r-project.org/softwaremap/trove_list.php?form_cat=353

[14]https://stat.ethz.ch/mailman/listinfo/R-SIG-Geo/

- The "Spatial.ly" [15] and "Kartograph" [16] webpages publish a variety of beautiful visualization examples.

---

[15]http://spatial.ly/r/
[16]http://kartograph.org/

Chapter 8

Thematic Maps: Proportional Symbol Mapping

The proportional symbol technique uses symbols of different sizes to represent data associated with areas or point locations, with circles being the most frequently used geometric symbol. The data and the size of symbols can be related through different types of scaling: mathematical scaling sizes areas of point symbols in direct proportion to the data; perceptual scaling corrects the mathematical scaling to account for visual understimation of larger symbols; and range grading, where data are grouped, and each class is represented with a single symbol size.

The most relevant packages used in this chapter are: sp and sf, and rgdal for reading and writing spatial data; maptools for label positioning; classInt for computing class intervals; gstat for spatial interpolation; mapview, plotKML, and rgl for interactive visualization.

8.1 Introduction

In this chapter we display data from a grid of sensors belonging to the Integrated Air Quality system of the Madrid City Council (Section 13.1) with circles as the proportional symbol, and range grading as the scaling method. The objective when using range grading is to discriminate between classes instead of estimating an exact value from a perceived symbol size. However, because human perception of symbol size is limited, it is always recommended to add a second perception channel to improve the discrimination task. Colors from a sequential palette will complement symbol size to encode the groups.

Two alternative are available to import and display the data:

1. Import the data with the `rgdal` package and the function `readOGR`, and display the information with the `sp` package and the `spplot` function (based on `lattice` graphics).

2. Import the data with the `sf` package and the function `st_read`, and display the information with the `ggplot2` package and the `geom_sf` function.

8.2 Proportional Symbol Mapping with `spplot`

The `spplot` method provided by the `sp` package, based on `xyplot` from the `lattice` package, is able to display `SpatialPointsDataFrame` objects, among others classes.

```
library(sp)
library(rgdal)

NO2sp <- readOGR(dsn = 'data/', layer = 'NO2sp')
```

A sequential palette is defined with `colorRampPalette`, with colors denoting the value of the variable (green for lower values of the contaminant, brown for intermediate values, and black for highest values).

```
airPal <- colorRampPalette(c('springgreen1', 'sienna3', 'gray5'))
    (5)
```

Both color and size can be combined in a unique graphical output because `spplot` accepts both of them (Figure 8.1).

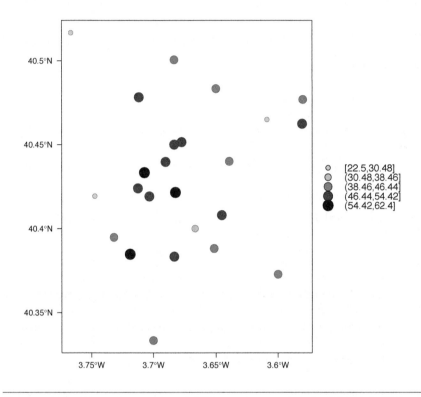

FIGURE 8.1: Annual average of NO_2 measurements in Madrid. Values are shown with different symbol sizes and colors for each class with the spplot function.

```
spplot(NO2sp["mean"],
       col.regions = airPal, ## Palette
       cex = sqrt(1:5), ## Size of circles
       edge.col = 'black', ## Color of border
       scales = list(draw = TRUE), ## Draw scales
       key.space = 'right') ## Put legend on the right
```

8.3 Proportional Symbol Mapping with ggplot

The ggplot2 package is able to display spatial point observations with the geom_sf function. This function understands the classes defined by the sf package, that provides the function st_read to read the data:

```
library(sf)

NO2sf <- st_read(dsn = 'data', layer = 'NO2sp')
## Create a categorical variable
NO2sf$Mean <- cut(NO2sf$mean, 5)

ggplot(data = NO2sf) +
    geom_sf(aes(size = Mean, fill = Mean),
            pch = 21, col = 'black') +
    scale_fill_manual(values = airPal) +
    theme_bw()
```

8.4   Optimal Classification to Improve Discrimination

Two main improvements can be added to Figure 8.1:

- Define classes dependent on the data structure (instead of the uniform distribution assumed with cut). A suitable approach is the classInterval function of the classInt package, which implements the Fisher-Jenks optimal classification algorithm[1]. This classification method seeks to reduce the variance within classes and maximize the variance between classes.

```
library(classInt)
## The number of classes is chosen between the Sturges and the
## Scott rules.
nClasses <- 5
intervals <- classIntervals(NO2sp$mean, n = nClasses, style = '
    fisher')
## Number of classes is not always the same as the proposed
    number
nClasses <- length(intervals$brks) - 1

op <- options(digits = 4)
tab <- print(intervals)
options(op)
```

---

[1]This classification method will be used in section 9.2 with a choropleth map.

FIGURE 8.2: Symbol sizes proposed by Borden Dent.

- Encode each group with a symbol size (circle area) such that visual discrimination among classes is enhanced. The next code uses the set of radii proposed in (Dent, Torguson, and Hodler 2008) (Figure 8.2). This set of circle sizes is derived from studies by Meihoefer (Meihoefer 1969). He derived a set of ten circle sizes that were easily and consistently discriminated by his subjects. The alternative proposed by Dent et al. improves the discrimination between some of the circles.

```
## Complete Dent set of circle radii (mm)
dent <- c(0.64, 1.14, 1.65, 2.79, 4.32, 6.22, 9.65, 12.95, 15.11)
## Subset for our dataset
dentAQ <- dent[seq_len(nClasses)]
## Link Size and Class: findCols returns the class number of each
## point; cex is the vector of sizes for each data point
idx <- findCols(intervals)
cexNO2 <- dentAQ[idx]
```

These two enhancements are included in Figure 8.3, which displays the categorical variable classNO2 (instead of mean) whose levels are the intervals previously computed with classIntervals. In addition, this figure includes an improved legend.

```
## spplot version
NO2sp$classNO2 <- factor(names(tab)[idx])

## Definition of an improved key with title and background
NO2key <- list(x = 0.99, y = 0.01, corner = c(1, 0),
               title = expression(NO[2]~~(paste(mu, plain(g))/m^3)),
               cex.title = 0.8, cex = 1,
               background = 'gray92')

pNO2 <- spplot(NO2sp["classNO2"],
               col.regions = airPal,
```

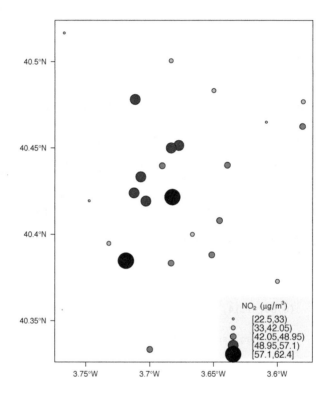

FIGURE 8.3: Annual average of NO_2 measurements in Madrid. Enhancement of Figure 8.1, using symbol sizes proposed by Borden Dent and an improved legend.

```
            cex = dentAQ * 0.8,
            edge.col = 'black',
            scales = list(draw = TRUE),
            key.space = NO2key)
pNO2
```

The ggplot version uses the same categorical variable, added to the NO2sf object.

```
## ggplot2 version
NO2sf$classNO2 <- factor(names(tab)[idx])
```

```
ggplot(data = NO2sf) +
    geom_sf(aes(size = classNO2, fill = classNO2),
            pch = 21, col = 'black') +
    scale_fill_manual(values = airPal) +
    scale_size_manual(values = dentAQ * 2) +
    xlab("") + ylab("") + theme_bw()
```

8.5 Spatial Context with Underlying Layers and Labels

The spatial distribution of the stations is better understood if we add underlying layers with information about the spatial context.

8.5.1 Static Image

A suitable method is to download data from a provider such as Google Maps[TM] or OpenStreetMap and transform it adequately. There are several packages that provide an interface to query several map servers: RGoogleMaps, OpenStreetMaps, and ggmap provide raster images from static maps obtained from Google Maps, Stamen, OpenStreetMap, etc.

Among these options, I have chosen the Stamen watercolor maps, and the ggmap package (Kahle and Wickham 2013). It is worth noting that these map tiles are published by Stamen Design under a Creative Commons licence CC BY-3.0 (Attribution). They produce these maps with data by OpenStreetMap also published under a Creative Commons licence BY-SA (Attribution - ShareAlike).

```
## Bounding box of data
madridBox <- bbox(NO2sp)
## Extend the limits to get a slightly larger map
madridBox <- t(apply(madridBox, 1,
                extendrange, f = 0.05))

library(ggmap)

madridGG <- get_map(c(madridBox),
                maptype = 'watercolor',
                source = 'stamen')
```

ggmap is designed to work with the ggplot2 package. Unfortunately, at the time of writing these pages, it is not able to work with geom_sf, so the SpatialPointsDataFrame object must be converted to a data.frame.

```
## ggmap with ggplot
NO2df <- as.data.frame(NO2sp)

ggmap(madridGG) +
    geom_point(data = NO2df,
                aes(coords.x1, coords.x2,
                    size = classNO2,
                    fill = classNO2),
                pch = 21, col = 'black') +
    scale_fill_manual(values = airPal) +
    scale_size_manual(values = dentAQ*2)
```

The result of get_map is only a raster object[2] with attributes. It can be added to a spplot graphic with the panel.ggmap function included in the sp package. This function is based on the grid.raster function. Previously, the SpatialPointsDataFrame must be transformed with sp-Transform because the Stamen maps use the Web Mercator projection[3]. The result, using the sp.layout argument, is shown in Figure 8.4.

```
## ggmap with spplot
## Project the data into the web mercator projection
NO2merc <- spTransform(NO2sp, CRS("+init=epsg:3857"))

## sp.layout definition
stamen <- list(panel.ggmap, ## Function that displays the object
                madridGG, ## Object to be displayed
                first = TRUE) ## This layout item will be drawn
                    before
                        ## the object displayed by spplot

spplot(NO2merc["classNO2"],
        col.regions = airPal,
        cex = dentAQ * 0.8,
        edge.col = 'black',
        sp.layout = stamen,
        scales = list(draw = TRUE),
        key.space = NO2key)
```

---

[2]Do not confuse a raster object with the Raster* objects of the raster package.
[3]https://epsg.io/3857

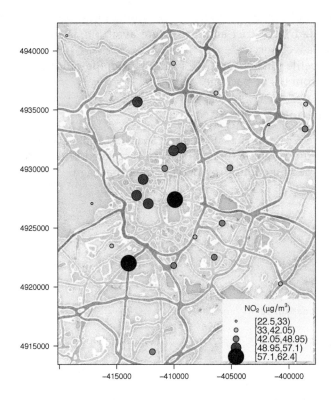

FIGURE 8.4: Annual average of NO_2 measurements in Madrid. Enhancement of Figure 8.3, using a Stamen map.

8.5.2 Vector Data

A major problem with the previous solution is that the user can neither modify the image nor use its content to produce additional information. A different approach is to use digital vector data (points, lines, and polygons). A popular format for vectorial data is the shapefile, commonly used by public and private providers to distribute information. For our example, the Madrid district and streets are available as shapefiles from the nomecalles web service[4].

A shapefile can be read with readOGR from the rgdal package, or with st_read from the sf package.

rgdal and spplot

The SpatialPolygonsDataFrame and SpatialLinesDataFrame objects produced by the readOGR function can be displayed with the sp.polygons and sp.lines functions provided by the sp package.

```
library(rgdal)

## nomecalles http://www.madrid.org/nomecalles/Callejero_
      madrid.icm
## Form at http://www.madrid.org/nomecalles/DescargaBDTCorte.icm

## Madrid districts
unzip('Distritos de Madrid.zip')
distritosMadrid <- readOGR('Distritos de Madrid/200001331.shp',
                    p4s = '+proj=utm +zone=30')
distritosMadrid <- spTransform(distritosMadrid,
                    CRS = CRS("+proj=longlat +ellps=WGS84"))

## Madrid streets
unzip('Callejero_ Ejes de viales.zip')
streets <- readOGR('Callejero_ Ejes de viales/call2011.shp',
            p4s = '+proj=utm +zone=30')
streetsMadrid <- streets[streets$CMUN=='079',]
streetsMadrid <- spTransform(streetsMadrid,
                    CRS = CRS("+proj=longlat +ellps=WGS84"))
```

These shapefiles can be included in the plot with the sp.layout mechanism accepted by spplot or with the layer and +.trellis functions from the latticeExtra package. The station codes are placed with this same

---

[4]http://www.madrid.org/nomecalles/

procedure using the `sp.pointLabel` function from the `maptools` package.
Figure 8.5 displays the final result.

```
library(maptools)
## Lists using the structure accepted by sp.layout, with the
     polygons,
## lines, and points, and their graphical parameters
spDistricts <- list('sp.polygons', distritosMadrid,
                fill = 'gray97', lwd = 0.3)
spStreets <- list('sp.lines', streetsMadrid,
                lwd = 0.05)
spNames <- list(sp.pointLabel, NO2sp,
             labels = substring(NO2sp$codEst, 7),
             cex = 0.6, fontfamily = 'Palatino')

## spplot with sp.layout version
spplot(NO2sp["classNO2"],
     col.regions = airPal,
     cex = dentAQ,
     edge.col = 'black',
     alpha = 0.8,
     ## Boundaries and labels overlaid
     sp.layout = list(spDistricts, spStreets, spNames),
     scales = list(draw = TRUE),
     key.space = NO2key)

## lattice with layer version
pNO2 +
    ## Labels *over* the original figure
    layer(sp.pointLabel(NO2sp,
                   labels = substring(NO2sp$codEst, 7),
                   cex = 0.8, fontfamily = 'Palatino')
        ) +
    ## Polygons and lines *below* (layer_) the figure
    layer_(
    {
       sp.polygons(distritosMadrid,
               fill = 'gray97',
               lwd = 0.3)
       sp.lines(streetsMadrid,
              lwd = 0.05)
    })
```

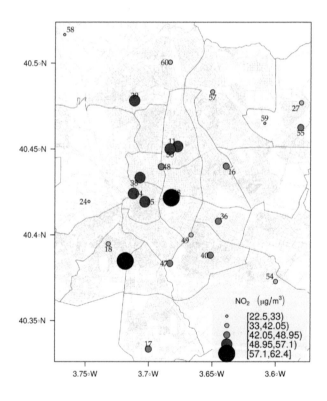

FIGURE 8.5: Annual average of NO_2 measurements in Madrid using shapefiles (lines and polygons) and text as geographical context.

sf and ggplot

The sf objects produced by the st_read function can be displayed with ggplot and geom_sf. The ggplot2 version of this figure uses geom_sf. The shapefiles must be read with the st_read function of the sf package.

```
library(sf)

## Madrid districts
distritosMadridSF <- st_read(dsn = 'Distritos de Madrid/',
                    layer = '200001331')
distritosMadridSF <- st_transform(distritosMadridSF,
                         crs = "+proj=longlat +ellps=WGS84")

## Madrid streets
streetsSF <- st_read(dsn = 'Callejero_ Ejes de viales/',
                    layer = 'call2011',
                    crs = '+proj=longlat +ellps=WGS84')

streetsMadridSF <- streetsSF[streetsSF$CMUN=='079',]
streetsMadridSF <- st_transform(streetsMadridSF,
                         crs = "+proj=longlat +ellps=WGS84")
```

8.6 Spatial Interpolation

The measurements at discrete points give limited information about the underlying process. It is quite common to approximate the spatial distribution of the measured variable with the interpolation between measurement locations. Selection of the optimal interpolation method is outside the scope of this book. The interested reader is referred to (Cressie and C. Wikle 2015) and (Bivand, E. J. Pebesma, and Gomez-Rubio 2013).

The following code illustrates an easy solution using inverse distance weighted (IDW) interpolation with the gstat package (E. J. Pebesma 2004) *only* for illustration purposes.

```
library(gstat)

## Sample 10^5 points locations within the bounding box of NO2sp
## using
## regular sampling
airGrid <- spsample(NO2sp, type = 'regular', n = 1e5)
## Convert the SpatialPoints object into a SpatialGrid object
gridded(airGrid) <- TRUE
```

```
## Compute the IDW interpolation
airKrige <- krige(mean ~ 1, NO2sp, airGrid)
```

The result is a `SpatialPixelsDataFrame` that can be displayed with spplot and combined with the previous layers and the measurement station points (Figure 8.6).

```
spplot(airKrige["var1.pred"], ## Variable interpolated
       col.regions = colorRampPalette(airPal)) +
   layer({ ## Overlay boundaries and points
      sp.polygons(distritosMadrid,
                  fill = 'transparent',
                  lwd = 0.3)
      sp.lines(streetsMadrid,
               lwd = 0.07)
      sp.points(NO2sp,
                pch = 21,
                alpha = 0.8,
                fill = 'gray50',
                col = 'black')
   })
```

8.7 Interactive Graphics

Now, let's suppose you need to know the median and standard deviation of the time series of a certain station. Moreover, you would like to view the photography of that station; or even better, you wish to visit its webpage for additional information. A frequent solution is to produce interactive graphics with tooltips and hyperlinks.

In this section we visit several approaches to create these products: the mapview package based on the `htmlwidgets` package; export to GeoJSON and KML formats; and 3D visualization with the `rgl` package.

8.7.1 mapview

The syntax of mapview[5] resembles the syntax of spplot. Its first argument is the spatial object with the information and the variable to be depicted is selected with the argument `zcol`. Moreover, the size of the points can

---

[5]The package mapview is able to work both with sp and sf objects. In this section the code works with sp objects, but would work without modification with sf objects.

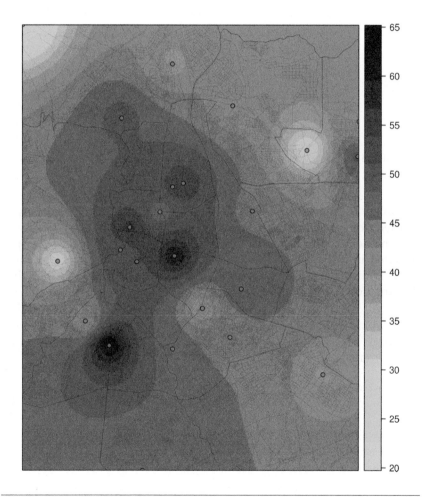

FIGURE 8.6: Kriging annual average of NO_2 measurements in Madrid.

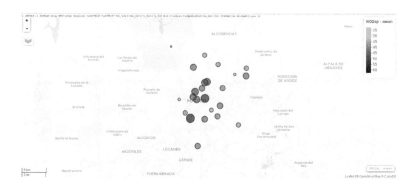

FIGURE 8.7: Snapshot of the interactive graphic produced with mapview depicting the annual average of NO_2 measurements in Madrid.

be linked to another variable with the argument cex, and their labels extracted from another variable with the argument label.

The next code produces an HTML page with an interactive graphic inserted in it (Figure 8.7). When the mouse is hovered over a point its label is displayed, and if the point is selected a tooltip with the whole information is deployed.

```
library(mapview)

pal <- colorRampPalette(c('springgreen1', 'sienna3', 'gray5'))
    (100)

mapview(NO2sp,
        zcol = "mean", ## Variable to display
        cex = "mean", ## Use this variable for the circle sizes
        col.regions = pal,
        label = NO2sp$Nombre,
        legend = TRUE)
```

Tooltips with images and graphs

The tooltip included in the previous graphic is very simple: only text displaying a table with information. This tooltip can be improved thanks to the popup argument and the popup* family of functions. For example, the next code creates an interactive graphic whose tooltips show an image

FIGURE 8.8: Snapshot of the interactive graphic produced with `mapview` with tooltips including images.

of the station (available in the `images` folder of the repository) using the `popupImage` function (Figure 8.8).

As an additional feature, the provider[6] of the background map is selected with the argument `map.type`.

```
img <- paste('images/', NO2sp$codEst, '.jpg', sep = '')

mapview(NO2sp,
        zcol = "mean",
        cex = "mean",
        col.regions = pal,
        label = NO2sp$Nombre,
        popup = popupImage(img, src = "local"),
        map.type = "Esri.WorldImagery",
        legend = TRUE)
```

A more sophisticated solution displays a scatterplot when a tooltip is deployed. The `popupGraph` function accepts a list of graphics and selects the one corresponding to the location selected by the user. This list is produced with the next code: first, the measurements time series is read and filtered; second, the stations code is extracted; finally, a loop with `lapply` creates a time series graphic for each station displaying the evolution of the measurements along the time period.

---

[6]The list of provider is available in `http://leaflet-extras.github.io/leaflet-providers/preview/`

```
## Read the time series
airQuality <- read.csv2('data/airQuality.csv')
## We need only NO2 data (codParam 8)
NO2 <- subset(airQuality, codParam == 8)
## Time index in a new column
NO2$tt <- with(NO2,
               as.Date(paste(year, month, day, sep = '-')))
## Stations code
stations <- unique(NO2$codEst)
## Loop to create a scatterplot for each station.
pList <- lapply(stations,
                function(i)
                    xyplot(dat ~ tt, data = NO2,
                           subset = (codEst == i),
                           type = 'l',
                           xlab = '', ylab = '')
                )
```

This list of graphics, pList, is provided to mapview through the popup argument with the function popupGraph (Figure 8.9).

```
mapview(NO2sp,
        zcol = "mean",
        cex = "mean",
        col.regions = pal,
        label = NO2sp$Nombre,
        popup = popupGraph(pList),
        map.type = "Esri.WorldImagery",
        legend = TRUE)
```

Synchronise multiple graphics

The mapview package recreates the small multiple technique (Sections 3.2 and 4.1) with the functions sync and latticeView. With them, multiple variables can be rendered simultaneously and synchronised together (with the sync function):

- if a panel is zoomed, all other panels will also zoom

- the mouse position in a panel is signaled with a red circle in the rest of panels.

The next code generates three graphics to view different variables of the NO2sp object using different values in zcol and cex. All of them are viewed and synchronised together with sync (Figure 8.10):

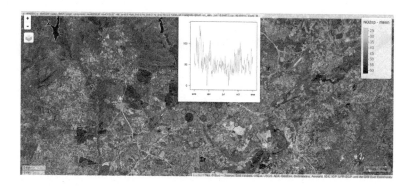

FIGURE 8.9: Snapshot of the interactive graphic produced with `mapview` with tooltips including time series graphics.

```
## Map of the average value
mapMean <- mapview(NO2sp, zcol = "mean", cex = "mean",
                col.regions = pal, legend = TRUE,
                map.types = "OpenStreetMap.Mapnik",
                label = NO2sp$Nombre)
## Map of the median
mapMedian <- mapview(NO2sp, zcol = "median", cex = "median",
                  col.regions = pal, legend = TRUE,
                  map.type = "Stamen.Watercolor",
                  label = NO2sp$Nombre)
## Map of the standard deviation
mapSD <- mapview(NO2sp, zcol = "sd", cex = "sd",
                col.regions = pal, legend = TRUE,
                map.type = "Esri.WorldImagery",
                label = NO2sp$Nombre)
## All together
sync(mapMean, mapMedian, mapSD, ncol = 3)
```

8.7.2 Export to Other Formats

A different approach is to use an external data viewer, due to its features or its large community of users. Two tools deserve to be mentioned: Geo-JSON rendered within GitHub repositories, and Keyhole Markup Language (KML) files imported in Google Earth™.

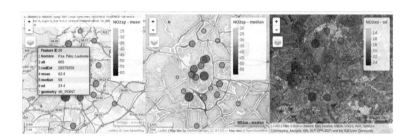

FIGURE 8.10: Snapshot of multiple interactive graphics produced with `mapview`.

GeoJSON and OpenStreetMap

GeoJSON is an open computer file format for encoding collections of simple geographical features along with their nonspatial attributes using JavaScript Object Notation (JSON). These files can be easily rendered within GitHub repositories. GitHub uses Leaflet[7] to represent the data and MapBox[8] with OpenStreetMap[9] for the underlying map data. Our `Spatial-PointsDataFrame` can be converted to a GeoJSON file with `writeOGR` from the `rgdal` package.

```
library(rgdal)
writeOGR(NO2sp, 'data/NO2.geojson', 'NO2sp', driver = 'GeoJSON')
```

Figure 8.11 shows a snapshot of the rendering of this GeoJSON file, available from the GitHub repository. There you can zoom on the map and click on the stations to display the data.

Keyhole Markup Language

Keyhole Markup Language (KML) is a file format to display geographic data within Internet-based, two-dimensional maps and three-dimensional Earth browsers. KML uses a tag-based structure with nested elements and attributes, and is based on the XML standard. KML became an international standard of the Open Geospatial Consortium in 2008. Google Earth was the first program able to view and graphically edit KML files, although Marble, an open-source project, also offers KML support.

---

[7] http://leafletjs.com/
[8] http://www.mapbox.com/
[9] https://www.openstreetmap.org/

FIGURE 8.11: NO_2 data in a GeoJSON file rendered within the GitHub repository.

There are several packages able to generate KML files. For example, the writeOGR function from the rgdal package can also write KML files:

```
library(rgdal)
writeOGR(NO2sp,
        dsn = 'NO2_mean.kml',
        layer = 'mean',
        driver = 'KML')
```

However, the plotKML package provides a simpler interface and includes a wide set of options:

```
library(plotKML)
plotKML(NO2sp["mean"], points_names = NO2sp$codEst)
```

Both functions produce a file that can be directly opened with Google Earth or Marble.

8.7.3 3D visualization

An alternative method is 3D visualization where the user can rotate or zoom the figure. This solution is available thanks to the rgl package, which provides functions for 3D interactive graphics.

```
library(rgl)
```

Previously, the `SpatialPointsDataFrame` object must be converted to a `data.frame`. The xyz coordinates will be the longitude, latitude, and altitude of each station.

```
## rgl does not understand Spatial* objects
NO2df <- as.data.frame(NO2sp)
```

The color of each point is determined by the corresponding class (Section 8.4), and the radius of each bubble depends on the mean value of the depicted variable.

```
## Color of each point according to its class
airPal <- colorRampPalette(c('springgreen1', 'sienna3', 'gray5'))
          (5)
colorClasses <- airPal[NO2df$classNO2]
```

A snapshot of this graphic is displayed in Figure 8.12.

```
plot3d(x = NO2df$coords.x1,
       y = NO2df$coords.x2,
       z = NO2df$alt,
       xlab = 'Longitude',
       ylab = 'Latitude',
       zlab = 'Altitude',
       type = 's',
       col = colorClasses,
       radius = NO2df$mean/10)
```

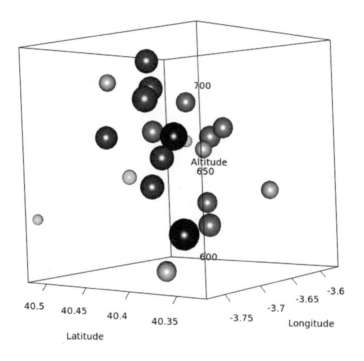

FIGURE 8.12: Snapshot of the interactive graphic produced with rgl.

Chapter 9

Thematic Maps: Choropleth Maps

A choropleth map shades regions according to the measurement of a variable displayed on the map. The choropleth map is an appropriate tool to visualize a variable uniformly distributed within each region, changing only at the region boundaries. This method performs correctly with homogeneous regions, both in size and shape. Next subjects are covered in this chapter: sequential and qualitative palettes; small multiples; class intervals; and interactive visualization.

The most relevant packages used in this chapter are: sp and sf, and rgdal for reading and writing spatial data; classInt for computing class intervals; RColorBrewer for color palettes; and mapview for interactive visualization.

9.1 Introduction

This chapter details how to create choropleth maps depicting the results of the 2016 Spanish general elections. The section 13.2 describes how to define a SpatialPolygonsDataFrame combining the data from a data.frame and the spatial information of the administrative boundaries from a shapefile[1].

As exposed in Chapter 8, two alternatives are available:

1. Import the data with the rgdal package and the function readOGR, and display the information with the sp package and the spplot function (based on lattice graphics).

2. Import the data with the sf package and the function st_read, and display the information with the ggplot2 package and the geom_sf function.

9.1.1 Read Data

The sp approach reads the files with readOGR to produce SpatialPolygons-DataFrame object. Because the coordinate reference system[2] is not stored in the files, it must be set with the p4s argument.

```
## sp approach
library(sp)
library(rgdal)

spMapVotes <- readOGR("data/spMapVotes.shp",
                 p4s = "+proj=utm +zone=30 +ellps=GRS80 +units=m
                     +no_defs")
```

This SpatialPolygonsDataFrame contains two main variables: which-Max, the name of the predominant political option, and pcMax, the percentage of votes obtained by this political option.

```
Object of class SpatialPolygonsDataFrame
Coordinates:
    min max
```

---

[1] The result is stored in the data folder, in a set of files named spMapVotes.*. You can visit it for details if you are interested in this procedure.

[2] The EPSG projection of the data is 25830, whose Proj4 definition is +proj=utm +zone=30 +ellps=GRS80 +units=m +no_defs. More information in http://spatialreference.org/ref/epsg/etrs89-utm-zone-30n/

```
x -13952 1127057
y 3903525 4859444
Is projected: TRUE
proj4string :
[+proj=utm +zone=30 +ellps=GRS80 +units=m +no_defs]
Data attributes:
    PROVMUN whichMax Max pcMax
 01001 : 1 ABS :2812 Min.  :  2.0 Min.  :21.33
 01002 : 1 C.s :  3 1st Qu.: 54.0 1st Qu.:31.68
 01003 : 1 OTH : 170 Median :  162.0 Median :35.64
 01004 : 1 PP :4212 Mean  : 1395.9 Mean :37.58
 01006 : 1 PSOE: 776 3rd Qu.: 636.5 3rd Qu.:41.27
 01008 : 1 UP  : 137 Max.  :696804.0 Max.  :94.74
 (Other):8104
```

The sf approach reads the files with st_read, and sets the coordinate reference system with st_crs, using directly the EPSG code[3].

```
## sf approach
library(sf)

sfMapVotes <- st_read("data/spMapVotes.shp")
st_crs(sfMapVotes) <- 25830
```

```
     SP_ID PROVMUN whichMax Max pcMax
 01001 : 1 01001 : 1 ABS :2812 Min.  :  2.0 Min.  :21.33
 01002 : 1 01002 : 1 C.s :  3 1st Qu.: 54.0 1st Qu.:31.68
 01003 : 1 01003 : 1 OTH : 170 Median :  162.0 Median :35.64
 01004 : 1 01004 : 1 PP :4212 Mean  : 1395.9 Mean :37.58
 01006 : 1 01006 : 1 PSOE: 776 3rd Qu.: 636.5 3rd Qu.:41.27
 01008 : 1 01008 : 1 UP  : 137 Max.  :696804.0 Max.  :94.74
 (Other):8104 (Other):8104
        geometry
MULTIPOLYGON :8110
epsg:25830 : 0
+proj=utm ...: 0
```

9.1.2 Province Boundaries

As a visual aid, the subsequent maps will be produced with the province boundaries superposed. These boundaries are encoded in the spain_-provinces shapefile. The spplot function that will generate the maps,

---

[3]http://spatialreference.org/ref/epsg/etrs89-utm-zone-30n/

offers the argument sp.layout to add additional layers to the map[4]. This argument should be a list: its first element is the function to display the layer, the second element is the object to be represented, and the rest of elements are arguments to the function defined in the first element.

```
## sp
provinces <- readOGR("data/spain_provinces.shp",
                p4s = "+proj=utm +zone=30 +ellps=GRS80 +units=m
                     +no_defs")

provinceLines <- list("sp.polygons", provinces, lwd = 0.1)

## sf
sfProvs <- st_read("data/spain_provinces.shp")
st_crs(sfProvs) <- 25830
```

9.2 Quantitative Variable

First, let's display pcMax, a quantitative variable increasing from low to high. This type of variables are well suited to sequential palettes, that communicate the progression from low to high with light colors associated to low values, and dark colors linked to high values. The well-known service ColorBrewer[5] provides several choices, available in R via the RColorBrewer package (Neuwirth 2014).

```
library(RColorBrewer)

## Number of intervals (colors)
N <- 6
## Sequential palette
quantPal <- brewer.pal(n = N, "Oranges")
```

The most common approach with choropleth maps, the classed choropleth, is to divide the data into classes. Although this method produces a filtered view of the data, it reduces the random noise in the information, and makes it easy to compare regions. A different alternative is the unclassed choropleth, where each unique data value gets a unique color. This approach is recommended to get a unfiltered view of the data and highlight overall geographic patterns of the variable.

---

[4]A similar result is obtained with the function layer of the latticeExtra package. The section 10.1 includes examples of this method.

[5]http://colorbrewer2.org

Figure 9.1 is an unclassed choropleth depicting the pcMax variable. It uses a palette created by interpolation with the colorRampPalette function.

```
## Number of cuts
ucN <- 1000
## Palette created with interpolation
ucQuantPal <- colorRampPalette(quantPal)(ucN)

## The polygons boundaries are not displayed thanks to col = '
    transparent'
spplot(spMapVotes["pcMax"],
       col.regions = ucQuantPal,
       cuts = ucN,
       ## Do not draw municipality boundaries
       col = 'transparent',
       ## Overlay province boundaries
       sp.layout = provinceLines)
```

9.2.1 Data Classification

It is evident in Figure 9.1 that the pcMax variable is concentrated in the 0.2-0.4 range. Figure 9.2 displays the density estimation of this variable grouping by the political option. This result suggests to use data classification.

```
ggplot(as.data.frame(spMapVotes),
       aes(pcMax,
           fill = whichMax,
           colour = whichMax)) +
    geom_density(alpha = 0.1) +
    theme_bw()
```

The number of data classes is the result of a compromise between information amount and map legibility. A general recommendation is to use three to seven classes, depending on the data.

On the other hand, there is a wide catalog of classification methods, and the classInt package implements most of them (previously used in Section 8.4). Figures 9.3 and 9.4 depict the empirical cumulative distribution function of pcMax with the intervals computed with the quantile method and the natural breaks method, a clustering method that seeks to reduce the variance within classes and maximize the variance between classes. As it can be inferred from the density estimation (Figure 9.2), the

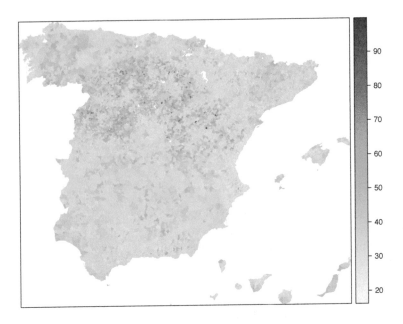

FIGURE 9.1: Quantitative choropleth map displaying the percentage of votes obtained by the predominant political option in each municipality in the 2016 Spanish general elections using a continuous color ramp (unclassed choropleth).

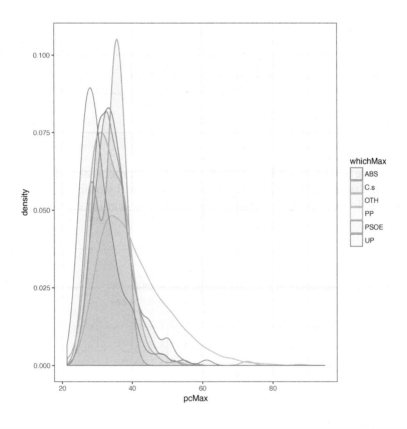

FIGURE 9.2: Density estimation of the predominant political option in each municipality in the 2016 Spanish general elections grouping by the political option.

natural breaks method is preferred in this example, because with the quantile method very different values will be assigned to the same class.

```
library(classInt)

## Compute intervals with the same number of elements
intQuant <- classIntervals(spMapVotes$pcMax,
                           n = N, style = "quantile")
## Compute intervals with the natural breaks algorithm
intFisher <- classIntervals(spMapVotes$pcMax,
                            n = N, style = "fisher")

plot(intQuant, pal = quantPal, main = "")

plot(intFisher, pal = quantPal, main = "")
```

Figure 9.5 is a classed choropleth with the natural breaks classification. It is produced with spplot displaying a categorical variable created with the function cut and the breaks computed with classIntervals.

```
## spplot solution

## Add a new categorical variable with cut, using the computed
   breaks
spMapVotes$pcMaxInt <- cut(spMapVotes$pcMax,
                           breaks = intFisher$brks)

spplot(spMapVotes["pcMaxInt"],
       col = 'transparent',
       col.regions = quantPal,
       sp.layout = provinceLines)

## sf and geom_sf
sfMapVotes$pcMaxInt <- cut(sfMapVotes$pcMax,
                           breaks = intFisher$brks)

ggplot(sfMapVotes) +
    ## Display the pcMaxInt variable...
    geom_sf(aes(fill = pcMaxInt),
            ## without drawing municipality boundaries
            color = "transparent") +
    scale_fill_brewer(palette = "Oranges") +
    ## And overlay provinces boundaries
    geom_sf(data = sfProvs,
```

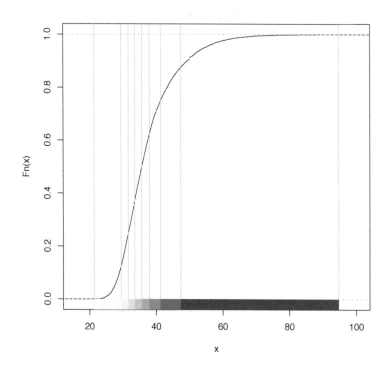

FIGURE 9.3: Quantile method for setting class intervals

```
        fill = 'transparent',
        ## but do not include them in the legend
        show.legend = FALSE) +
theme_bw()
```

9.3 Qualitative Variable

On the other hand, whichMax is a categorical value with four levels: the main parties (PP, PSOE, UP, Cs), the abstention results (ABS), and the rest of the parties (OTH).

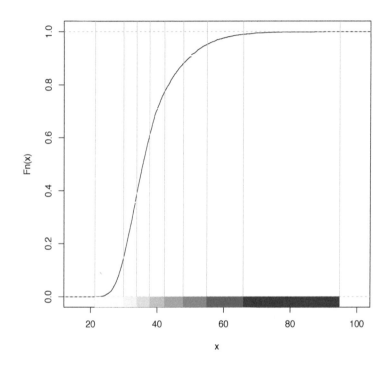

FIGURE 9.4: Natural breaks method for setting class intervals

```
ABS  C.s  OTH   PP PSOE   UP
2812    3  170 4212  776  137
```

Figure 9.6 encodes these levels with a qualitative palette from Color-Brewer.

```
classes <- levels(factor(spMapVotes$whichMax))
nClasses <- length(classes)

qualPal <- brewer.pal(nClasses, "Dark2")

## spplot solution
spplot(spMapVotes["whichMax"],
```

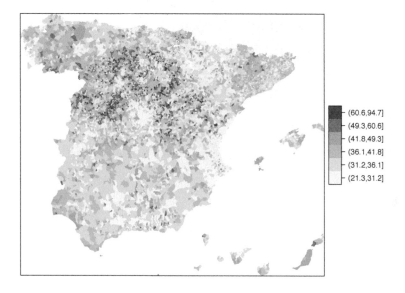

FIGURE 9.5: Quantitative choropleth map displaying the percentage of votes obtained by the predominant political option in each municipality in the 2016 Spanish general elections using a classification (classed choropleth).

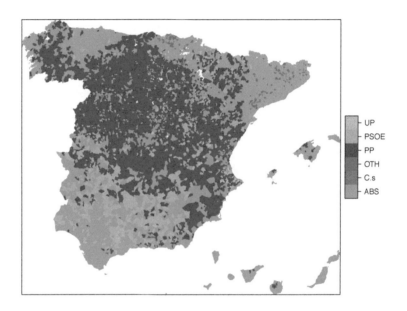

FIGURE 9.6: Categorical choropleth map displaying the name of the predominant political option in each municipality in the 2016 Spanish general
elections.

```
        col.regions = qualPal,
        col = 'transparent',
        sp.layout = provinceLines)

## geom_sf solution
ggplot(sfMapVotes) +
    geom_sf(aes(fill = whichMax),
            color = "transparent") +
    scale_fill_brewer(palette = 'Dark2') +
    geom_sf(data = sfProvs,
            fill = 'transparent',
```

```
        show.legend = FALSE) +
    theme_bw()
```

9.4 Small Multiples with Choropleth Maps

Both the quantitative and qualitative variables can be combined using the small multiples technique (Sections 3.2 and 4.1) (Tufte 1990): multiple maps displayed all at once to compare the differences between them. The next code produce a matrix of maps, with a map for each political option defined by the categorical variable whichMax. The spplot function provides a formula argument to divide the data into panels. However, its usage is not well documented and cannot be recommended. Instead, the ggplot approach is easy to use thanks to the facet_wrap function. The result is displayed in Figure 9.7.

```
ggplot(sfMapVotes) +
    geom_sf(aes(fill = pcMaxInt),
            color = "transparent") +
    ## Define the faceting using two rows
    facet_wrap(~whichMax, nrow = 2) +
    scale_fill_brewer(palette = "Oranges") +
    geom_sf(data = sfProvs,
            fill = 'transparent',
            size = 0.1,
            show.legend = FALSE) +
    theme_bw()
```

9.5 Bivariate Map

Following the inspiring example of the infographic from the *New York Times*, we will combine the choropleth maps of both variables to produce a bivariate map[6]: the hue of each polygon will be determined by the name of the predominant option (whichMax) but the transparency will vary according to the percentage of votes (pcMax).

In previous sections, we use six intervals to represent the quantitative variable pcMax. However, in this case we must reduce this number: in

---

[6]Although bivariate maps are generally used to display the relationship between two variables, they can also be used to display one variable and its uncertainty. More information about visualizing uncertainty with maps can be found in (Lucchesi and C. K. Wikle 2017) and the package VizU (https://github.com/pkuhnert/VizU).

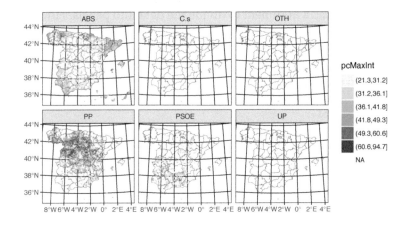

FIGURE 9.7: Small multiple choropleth maps of the Spanish general elections results. Each map shows the results of a political option in each municipality.

order to improve the map legibility, each ramp has only three steps. Thus, the bivariate legend will be composed of eighteen colors.

Next code creates a bidimensional palette with a loop that produces a color ramp for each member of the qualitative palette. Each ramp begins in the original color of the palette, and ends in a lighter color with its transparency fixed to 40%.

```
## Number of intervals.
N <- 3
## Loop to create a bidimensional palette
multiPal <- sapply(1:nClasses, function(i)
{
    colorAlpha <- adjustcolor(qualPal[i], alpha = 0.4)
    colorRampPalette(c(qualPal[i], colorAlpha), alpha = TRUE)(N)
})

## Define the intervals
intFisher <- classIntervals(spMapVotes$pcMax,
                        n = N, style = "fisher")
## ... and create a categorical variable with them
spMapVotes$pcMaxInt <- cut(spMapVotes$pcMax,
                        breaks = intFisher$brks)
```

With this multivariate palette we can produce a list of maps extracting the polygons according to each class of the qualitative variable, and filling with the appropiate color from this palette. The resulting list of `trellis` objects can be combined with `Reduce` and the `+.trellis` function of the `latticeExtra` and produce a `trellis` object.

```
pList <- lapply(1:nClasses, function(i){
    ## Only those polygons corresponding to a level are selected
    mapClass <- subset(spMapVotes,
                    whichMax == classes[i])
    ## Palette
    pal <- multiPal[, i]
    ## Produce the graphic
    pClass <- spplot(mapClass, "pcMaxInt",
                col.regions = pal,
                col = 'transparent',
                colorkey = FALSE)
})
names(pList) <- classes
p <- Reduce('+', pList)
```

The bidimensional legend of this graphic is produced with grid.raster, a function of the grid package, able to display a color matrix (line 13). The axis of the color matrix are created with grid.text (lines 19 and 30).

```
1    library(grid)
2
3    legend <- layer(
4    {
5        ## Position of the legend
6        x0 <- 1000000
7        y0 <- 4200000
8        ## Width of the legend
9        w <- 120000
10       ## Height of the legend
11       h <- 100000
12       ## Colors
13       grid.raster(multiPal, interpolate = FALSE,
14                       x = unit(x0, "native"),
15                       y = unit(y0, "native"),
16                   width = unit(w, "native"),
17                   height = unit(h, "native"))
18       ## x-axis (qualitative variable)
19       grid.text(classes,
20               x = unit(seq(x0 - w * (nClasses -1)/(2*nClasses),
21                           x0 + w * (nClasses -1)/(2*nClasses),
22                           length = nClasses),
23                   "native"),
24               y = unit(y0 + h/2, "native"),
25               just = "bottom",
26               rot = 10,
27               gp = gpar(fontsize = 4))
28       ## y-axis (quantitative variable)
29       Ni <- length(intervals)
30       grid.text(intervals,
31               x = unit(x0 + w/2, "native"),
32               y = unit(seq(y0 - h * (Ni -1)/(2*Ni),
33                           y0 + h * (Ni -1)/(2*Ni),
34                           length = Ni),
35                   "native"),
36               just = "left",
37               gp = gpar(fontsize = 6))
38   })
```

Figure 9.8 displays the result.

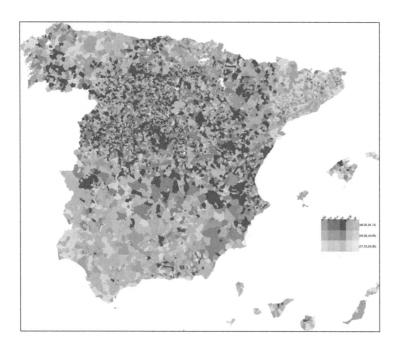

FIGURE 9.8: Bidimensional choropleth map of the Spanish general elections results. The map shows the result of the most voted option in each municipality.

```
## Main plot
p + legend
```

9.6 Interactive Graphics

The package mapview was used in section 8.7.1 to produce interactive proportional symbol maps. In this section this package creates interactive choropleth maps.

```
library(mapview)
```

FIGURE 9.9: Snapshot of the interactive quantitative choropleth map produced with mapview.

This package is able to work both with sp and with sf. In this section we use the sf package to read the data[7].

```
sfMapVotes0 <- st_read("data/spMapVotes0.shp")
st_crs(sfMapVotes0) <- 25830
```

Figures 9.9 and 9.10 show the snapshots of the interactive choropleth maps of pcMax and whichMax, respectively. These maps are produced with the next code.

```
## Quantitative variable, pcMax
mapView(sfMapVotes0,
        zcol = "pcMax", ## Choose the variable to display
        legend = TRUE,
        col.regions = quantPal)


## Qualitative variable, whichMax
mapView(sfMapVotes0,
        zcol = "whichMax",
```

---

[7]In previous sections the spatial object included a modification to the original shapefile in order to display the Canarian islands in the right bottom corner of the maps. This modification is not needed with mapview, so st_read imports the shapefile spMapVotes0 (Section 13.2).

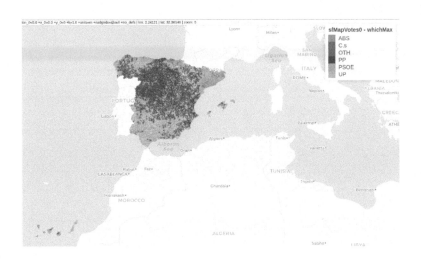

FIGURE 9.10: Snapshot of the interactive qualitative choropleth map produced with `mapview`.

```
legend = TRUE,
col.regions = qualPal)
```

Chapter 10

Thematic Maps: Raster Maps

A raster data structure is a matrix of cells organized into rows and columns where each cell contains a value representing information, such as temperature, altitude, population density, land use, etc. This chapter describes how to display a raster with two different data sets: CM-SAF solar irradiation rasters will illustrate the use of quantitative data, and land cover and population data from the NEO-NASA project will exemplify the display of categorical data and multivariate rasters.

Main subjects covered in this chapter are: sequential, qualitative, and diverging palettes; class intervals and visual discrimination; hill shading; 3D and interactive visualization.

The most relevant packages used in this chapter are: `raster` for reading raster data; `rasterVis` for visualization of raster data; `maps`, `mapdata`, and `maptools` for boundaries lines; `classInt` for computing class intervals; `mapview` for interactive maps.

10.1   Quantitative Data

As an example of quantitative data, this section displays the distribution of annual solar irradiation over the Iberian peninsula using the estimates from CM SAF. Read Chapter 13 for details about this dataset.

```
library(raster)
library(rasterVis)
```

```
SISav <- raster('data/SISav')
```

The RasterLayer object of annual averages of solar irradiation estimated by CM SAF can be easily displayed with the levelplot method of the rasterVis package. Figure 10.1 illustrates this raster with marginal graphics to show the column (longitude) and row (latitude) summaries of the RasterLayer object. The summary is computed with the function defined by FUN.margin (which uses mean as the default value).

```
levelplot(SISav)
```

Although the solar irradiation distribution reveals the physical structure of the region, it is recommended to add the geographic context with a layer of administrative boundaries (Figure 10.2).

```
library(maps)
library(mapdata)
library(maptools)
## Extent of the Raster object
ext <- as.vector(extent(SISav))
## Retrieve the boundaries restricted to this extent
boundaries <- map('worldHires',
                  xlim = ext[1:2], ylim = ext[3:4],
                  plot = FALSE)
## Convert the result to a SpatialLines with the projection of
    the
## Raster object
boundaries <- map2SpatialLines(boundaries,
                    proj4string = CRS(projection(SISav)))

## Display the Raster object ...
levelplot(SISav) +
    ## ... and overlay the SpatialLines object
    layer(sp.lines(boundaries,
               lwd = 0.5))
```

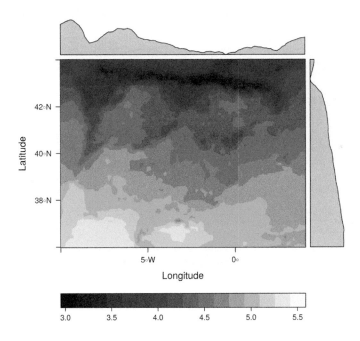

FIGURE 10.1: Annual average of solar radiation displayed with a sequential palette.

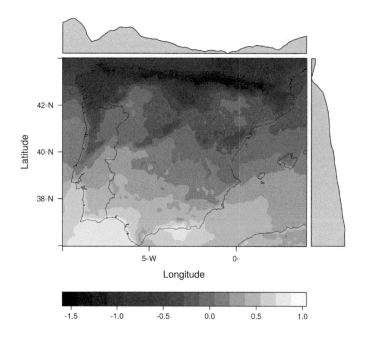

FIGURE 10.2: Annual average of solar radiation with administrative boundaries.

10.1.1   Hill Shading

A frequent method to improve the display of meteorological rasters is the hill shading or shaded relief technique, a method of representing relief on a map by depicting the shadows that would be cast by high ground if light comes from a certain sun position (Figure 10.3). The hill shade layer can be computed from the slope and aspect layers derived from a Digital Elevation Model (DEM). This layer will underlay the DEM raster, which will be printed using semitransparency.

The procedure is as follows:

- Download a Digital Elevation Model (DEM) from the DIVA-GIS service.

```
old <- setwd(tempdir())
download.file('http://biogeo.ucdavis.edu/data/diva/msk_alt/ESP_
    msk_alt.zip', 'ESP_msk_alt.zip')
unzip('ESP_msk_alt.zip', exdir = '.')

DEM <- raster('ESP_msk_alt')
```

- Compute the hill shade raster with `terrain` and `hillShade` from raster with a certain angle and direction.

```
slope <- terrain(DEM, 'slope')
aspect <- terrain(DEM, 'aspect')
hs <- hillShade(slope = slope, aspect = aspect,
                angle = 60, direction = 45)

setwd(old)
```

- Combine the result with the previous map using semitransparency.

```
## hillShade theme: gray colors and semitransparency
hsTheme <- GrTheme(regions = list(alpha = 0.5))

levelplot(SISav,
        par.settings = YlOrRdTheme,
        margin = FALSE, colorkey = FALSE) +
    ## Overlay the hill shade raster
    levelplot(hs, par.settings = hsTheme, maxpixels = 1e6) +
    ## and the countries boundaries
    layer(sp.lines(boundaries, lwd = 0.5))
```

153

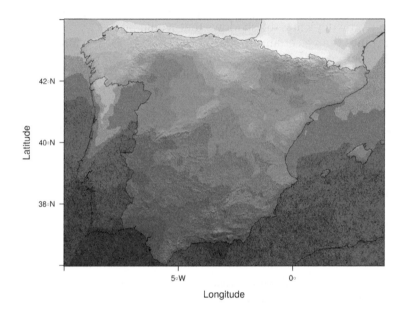

FIGURE 10.3: Hill shading of annual average of solar radiation.

10.1.2 Diverging Palettes

Next, instead of displaying the absolute values of each cell, we will analyze the differences between each cell and the global average value. This average is computed with the cellStats function and substracted from the original RasterLayer.

```
meanRad <- cellStats(SISav, 'mean')
SISav <- SISav - meanRad
```

Figure 10.4 displays the relation between these scaled values and latitude (y), with five different groups defined by the longitude (cut(x, 5)). It is evident that larger irradiation values are associated with lower lati-

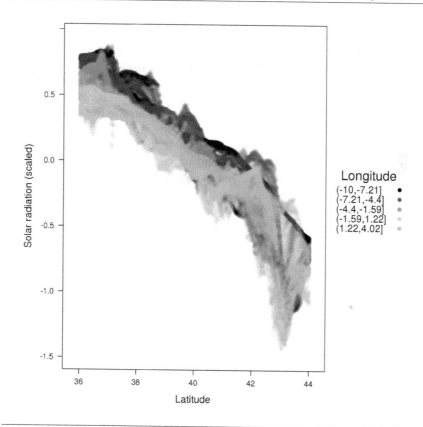

FIGURE 10.4: Relation between scaled annual average radiation and latitude for several longitude groups.

tudes. However, there is no such clear relation between irradiation and longitude.

```
xyplot(layer ~ y, data = SISav,
       groups = cut(x, 5),
       par.settings = rasterTheme(symbol = plinrain(n = 5,
                                                     end = 200)),
       xlab = 'Latitude', ylab = 'Solar radiation (scaled)',
       auto.key = list(space = 'right',
                       title = 'Longitude',
                       cex.title = 1.3))
```

Numerical information ranging in an interval including a neutral value is commonly displayed with diverging palettes. These palettes represent neutral classes with light colors, while low and high extremes of the data range are highlighted using dark colors with contrasting hues. I use the Purple-Orange palette from ColorBrewer with purple for positive values and orange for negative values. In order to underline the position of the interval containing zero, the center color of this palette is substituted with pure white. The resulting palette is displayed in Figure 10.5 with the custom showPal function. The corresponding correspondent raster map produced with this palette is displayed in Figure 10.6. Although extreme positive and negative values can be easily discriminated, the zero value is not associated with white because the data range is not symmetrical around zero.

```
divPal <- brewer.pal(n = 9, 'PuOr')
divPal[5] <- "#FFFFFF"

showPal <- function(pal)
{
   N <- length(pal)
   image(1:N, 1, as.matrix(1:N), col = pal,
         xlab = '', ylab = '',
         xaxt = "n", yaxt = "n",
         bty = "n")
}

showPal(divPal)

divTheme <- rasterTheme(region = divPal)

levelplot(SISav, contour = TRUE, par.settings = divTheme)
```

The solution is to connect the symmetrical color palette with the asymmetrical data range. The first step is to create a set of breaks such that the zero value is the center of one of the intervals.

```
rng <- range(SISav[])
## Number of desired intervals
nInt <- 15
## Increment corresponding to the range and nInt
inc0 <- diff(rng)/nInt
## Number of intervals from the negative extreme to zero
n0 <- floor(abs(rng[1])/inc0)
```

FIGURE 10.5: Purple-Orange diverging palette using white as middle color.

```
## Update the increment adding 1/2 to position zero in the center
     of an interval
inc <- abs(rng[1])/(n0 + 1/2)
## Number of intervals from zero to the positive extreme
n1 <- ceiling((rng[2]/inc - 1/2) + 1)
## Collection of breaks
breaks <- seq(rng[1], by = inc, length= n0 + 1 + n1)
```

The next step is to compute the midpoints of each interval. These points represent the data belonging to each interval, and their value will be connected with a color of the palette.

```
## Midpoints computed with the median of each interval
idx <- findInterval(SISav[], breaks, rightmost.closed = TRUE)
mids <- tapply(SISav[], idx, median)
## Maximum of the absolute value both limits
mx <- max(abs(breaks))
```

A simple method to relate the palette and the intervals is with a straight line such that a point is defined by the absolute maximum value, ((mx, 1)), and another point by zero, ((0, 0.5)). Why are we using the interval [0, 1] as the y-coordinate of this line, and why is 0.5 the result of zero? The reason is that the input of the break2pal function will be the result of colorRamp, a function that creates another interpolating function which maps colors with values between 0 and 1. Therefore, a new palette is created, extracting colors from the original palette, such that the central color (white)

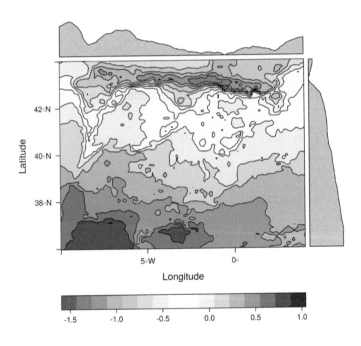

FIGURE 10.6: Asymmetric raster data (scaled annual average irradiation) displayed with a symmetric diverging palette.

FIGURE 10.7: Modified diverging palette related with the asymmetrical raster data.

is associated with the interval containing zero. This palette is displayed in Figure 10.7.

The raster map produced with this new palette is displayed in Figure 10.8. Now zero is clearly associated with the white color.

```
break2pal <- function(x, mx, pal){
    ## x = mx gives y = 1
    ## x = 0 gives y = 0.5
    y <- 1/2*(x/mx + 1)
    rgb(pal(y), maxColorValue = 255)
}

## Interpolating function that maps colors with [0, 1]
## rgb(divRamp(0.5), maxColorValue=255) gives "#FFFFFF" (white)
divRamp <- colorRamp(divPal)
## Diverging palette where white is associated with the interval
## containing the zero
pal <- break2pal(mids, mx, divRamp)
showPal(pal)

levelplot(SISav,
          par.settings = rasterTheme(region = pal),
          at = breaks, contour = TRUE)
```

It is interesting to note two operations carried out internally by the lattice package. First, the custom.theme function (used by rasterTheme)

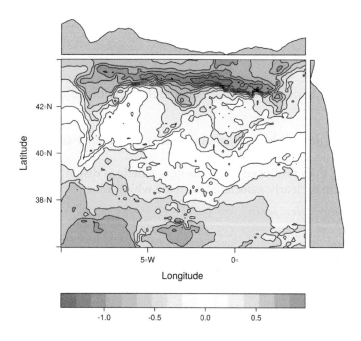

FIGURE 10.8: Asymmetric raster data (scaled annual average irradiation) displayed with a modified diverging palette.

creates a new palette with 100 colors using `colorRampPalette` to interpolate the palette passed as an argument. Second, the `level.colors` function makes the arrangement between intervals and colors. If this function receives more colors than intervals, it chooses a subset of the palette disregarding some of the intermediate colors. Therefore, because this function will receive 100 colors from `par.settings`, it is difficult to control exactly which colors of our original palette will be represented.

An alternative way for finer control is to fill the `regions$col` component of the theme with our palette after it has been created (Figure 10.9).

```
divTheme <- rasterTheme(regions = list(col = pal))

levelplot(SISav,
          par.settings = divTheme,
          at = breaks,
          contour = TRUE)
```

A final improvement to this map is to compute the intervals using a classification algorithm with the `classInt` package. With this approach it is likely that zero will not be perfectly centered in its corresponding interval. The remaining code is exactly the same as above, replacing the `breaks` vector with the result of the `classIntervals` function. Figure 10.10 displays the result.

```
library(classInt)

cl <- classIntervals(SISav[], style = 'kmeans')
breaks <- cl$brks

## Repeat the procedure previously exposed, using the 'breaks'
    vector
## computed with classIntervals
idx <- findInterval(SISav[], breaks, rightmost.closed = TRUE)
mids <- tapply(SISav[], idx, median)

mx <- max(abs(breaks))
pal <- break2pal(mids, mx, divRamp)

## Modify the vector of colors in the 'divTheme' object
divTheme$regions$col <- pal

levelplot(SISav,
          par.settings = divTheme,
          at = breaks,
          contour = TRUE)
```

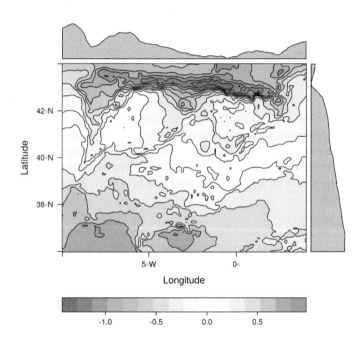

FIGURE 10.9: Same as Figure 10.8 but colors are assigned directly to the regions$col component of the theme.

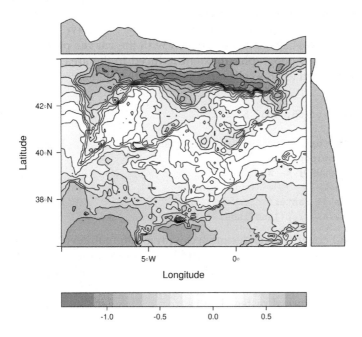

FIGURE 10.10: Same as Figure 10.9 but defining intervals with the optimal classification method.

10.2 Categorical Data

Land cover is the observed physical cover on the Earth's surface. A set of seventeen different categories is commonly used. Using satellite observations, it is possible to map where on Earth each of these seventeen land surface categories can be found and how these land covers change over time.

This section illustrates how to read and display rasters with categorical information using information from the NEO-NASA project. Read Chapter 13 for details about these datasets.

After the land cover and population density files have been downloaded, two `RasterLayers` can be created with the `raster` package. Both files are read, their geographical extent reduced to the area of India and China, and cleaned (99999 cells are replaced with `NA`).

```
## China and India
ext <- extent(65, 135, 5, 55)

pop <- raster('data/875430rgb-167772161.0.FLOAT.TIFF')
pop <- crop(pop, ext)
pop[pop==99999] <- NA

landClass <- raster('data/241243rgb-167772161.0.TIFF')
landClass <- crop(landClass, ext)
```

Each land cover type is designated with a different key: the sea is labeled with 0; forests with 1 to 5; shrublands, grasslands, and wetlands with 6 to 11; agriculture and urban lands with 12 to 14; and snow and barren with 15 and 16. These four groups (sea is replaced by `NA`) will be the levels of the categorical raster. The `raster` package includes the `ratify` method to define a layer as categorical data, filling it with integer values associated to a Raster Attribute Table (RAT).

```
landClass[landClass %in% c(0, 254)] <- NA
## Only four groups are needed:
## Forests: 1:5
## Shrublands, etc: 6:11
## Agricultural/Urban: 12:14
## Snow: 15:16
landClass <- cut(landClass, c(0, 5, 11, 14, 16))
## Add a Raster Attribute Table and define the raster as
##    categorical data
landClass <- ratify(landClass)
```

```
## Configure the RAT: first create a RAT data.frame using the
## levels method; second, set the values for each class (to be
## used by levelplot); third, assign this RAT to the raster
## using again levels
rat <- levels(landClass)[[1]]
rat$classes <- c('Forest', 'Land', 'Urban', 'Snow')
levels(landClass) <- rat
```

This categorical raster can be displayed with the `levelplot` method of the `rasterVis` package. Previously, a theme is defined with the background color set to `lightskyblue1` to display the sea areas (filled with NA values), and the region palette is defined with adequate colors (Figure 10.11).

```
qualPal <- c('palegreen4', # Forest
        'lightgoldenrod', # Land
        'indianred4', # Urban
        'snow3') # Snow

qualTheme <- rasterTheme(region = qualPal,
                panel.background = list(col = 'lightskyblue1')
                )

levelplot(landClass, maxpixels = 3.5e5,
        par.settings = qualTheme)
```

Let's explore the relation between the land cover and population density rasters. Figure 10.12 displays this latter raster using a logarithmic scale, defined with `zscaleLog`.

```
pPop <- levelplot(pop, zscaleLog = 10,
                par.settings = BTCTheme,
                maxpixels = 3.5e5)
pPop
```

Both rasters can be joined together with the `stack` method to create a new `RasterStack` object. Figure 10.13 displays the distribution of the logarithm of the population density associated to each land class.

```
## Join the RasterLayer objects to create a RasterStack object.
s <- stack(pop, landClass)
names(s) <- c('pop', 'landClass')

densityplot(~log10(pop), ## Represent the population
        groups = landClass, ## grouping by land classes
```

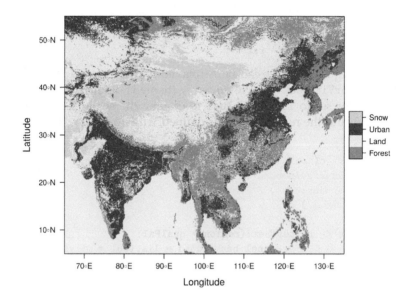

FIGURE 10.11: Land cover raster (categorical data).

```
data = s,
## Do not plot points below the curves
plot.points = FALSE)
```

10.3 🐱Bivariate Legend

We can reproduce the code used to create the multivariate choropleth (Section 9.5) using the levelplot function from the rasterVis package. Again, the result is a list of trellis objects. Each of these objects is the representation of the population density in a particular land class.

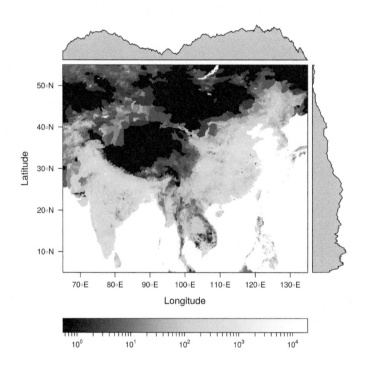

FIGURE 10.12: Population density raster.

```
classes <- rat$classes
nClasses <- length(classes)

logPopAt <- c(0, 0.5, 1.85, 4)

nIntervals <- length(logPopAt) - 1

multiPal <- sapply(1:nClasses, function(i)
{
    colorAlpha <- adjustcolor(qualPal[i], alpha = 0.4)
    colorRampPalette(c(qualPal[i],
                colorAlpha),
```

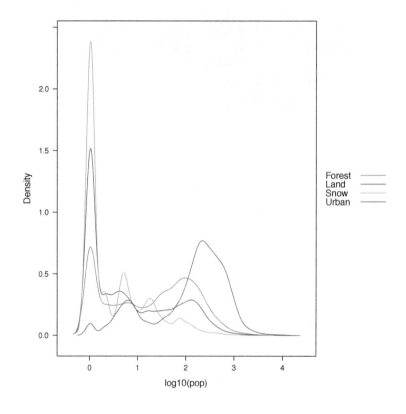

FIGURE 10.13: Distribution of the logarithm of the population density associated to each land class.

```
                    alpha = TRUE)(nIntervals)
})

pList <- lapply(1:nClasses, function(i){
    landSub <- landClass
    ## Those cells from a different land class are set to NA...
    landSub[!(landClass==i)] <- NA
    ## ... and the resulting raster masks the population raster
    popSub <- mask(pop, landSub)
    ## Palette
    pal <- multiPal[, i]

    pClass <- levelplot(log10(popSub),
                    at = logPopAt,
                    maxpixels = 3.5e5,
                    col.regions = pal,
                    colorkey = FALSE,
                    margin = FALSE)
})
```

The +.trellis function of the latticeExtra package with Reduce superposes the elements of this list and produces a trellis object.

```
p <- Reduce('+', pList)
```

The legend is created with grid.raster and grid.text, following the same procedure exposed in section 9.5.

```
library(grid)

legend <- layer(
{
    ## Center of the legend (rectangle)
    x0 <- 125
    y0 <- 22
    ## Width and height of the legend
    w <- 10
    h <- w / nClasses * nIntervals
    ## Legend
    grid.raster(multiPal, interpolate = FALSE,
                x = unit(x0, "native"),
                y = unit(y0, "native"),
            width = unit(w, "native"))
    ## Axes of the legend
    ## x-axis (qualitative variable)
```

```
grid.text(classes,
        x = unit(seq(x0 - w * (nClasses -1)/(2*nClasses),
                     x0 + w * (nClasses -1)/(2*nClasses),
                     length = nClasses),
                "native"),
        y = unit(y0 + h/2, "native"),
        just = "bottom",
        rot = 10,
        gp = gpar(fontsize = 6))
    ## y-axis (quantitative variable)
    yLabs <- paste0("[",
            paste(logPopAt[-nIntervals],
                logPopAt[-1], sep = ","),
            "]")
    grid.text(yLabs,
        x = unit(x0 + w/2, "native"),
        y = unit(seq(y0 - h * (nIntervals -1)/(2*nIntervals),
                     y0 + h * (nIntervals -1)/(2*nIntervals),
                     length = nIntervals),
                "native"),
        just = "left",
        gp = gpar(fontsize = 6))

})
```

Figure 10.14 displays the result with the legend.

```
p + legend
```

10.4   Interactive Graphics

10.4.1   3D Visualization

An alternative method for a DEM is 3D visualization where the user can rotate or zoom the figure. This solution is available thanks to the rgl package, which provides functions for 3D interactive graphics. The plot3D function in the rasterVis package is a wrapper to this package for Raster-Layer objects.

```
plot3D(DEM, maxpixels = 5e4)
```

The output scene can be exported to several formats such as STL with writeSTL, a format commonly used in 3D printing, or WebGL with writeWe-bGL to be rendered in a browser (Figure 10.15).

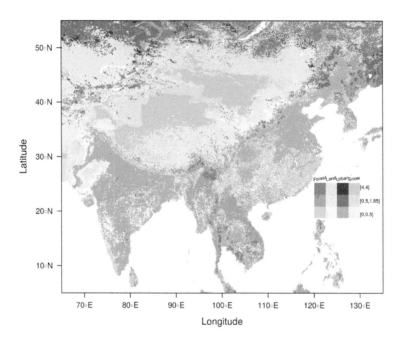

FIGURE 10.14: Population density for each land class (multivariate raster).

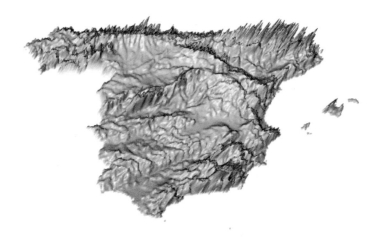

FIGURE 10.15: 3D visualization of a Digital Elevation Model using the WebGL format.

```
## Dimensions of the window in pixels
par3d(viewport = c(0, 30, ## Coordinates of the lower left corner
             250, 250)) ## Width and height

writeWebGL(filename = 'docs/images/rgl/DEM.html',
        width = 800)
```

10.4.2 Mapview

The package mapview is able to work with Raster* objects. Thus, the SISav object can be easily displayed in a interactive map with next code. However, it must be noted that, unlike with vector data (Sections 8.7.1 and 9.6), the interactivity of this map is restricted to zoom and movement. The mouse hovering or click does not produce any result.

```
library(mapview)

mvSIS <- mapview(SISav, legend = TRUE)
```

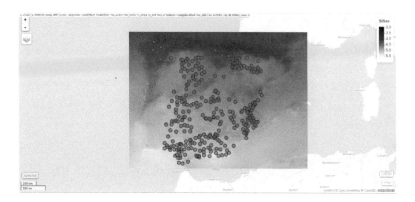

FIGURE 10.16: Snapshot of the interactive map produced with `mapview` combining a `RasterLayer` and a `SpatialPointsDataFrame` objects.

This map can be improved with another layer of information, the name and location of the meteorological stations of the SIAR network. This information is stored in the file `SIAR.csv`. Next code reads this file and produces a `SpatialPointsDataFrame` object.

```
SIAR <- read.csv("data/SIAR.csv")

spSIAR <- SpatialPointsDataFrame(coords = SIAR[, c("lon", "lat")
    ],
                    data = SIAR,
                    proj4str = CRS(projection(SISav)))
```

This object, as shown in section 8.7.1, can also be displayed with `mapview`. The resulting map is reactive to mouse hovering and click.

```
mvSIAR <- mapview(spSIAR,
            label = spSIAR$Estacion)
```

Both layers of information can be combined with the + operator. Figure 10.16 shows a snapshot of this interactive map.

```
mvSIS + mvSIAR
```

Chapter 11

Vector Fields

Many objects in our natural environment exhibit directional features that are naturally represented by vector data. Vector fields, commonly found in science and engineering, describe the spatial distribution of a vector variable such as fluid flow or electromagnetic forces. A suitable visualization method has to display both the magnitude and the direction of the vectors at any point.

This chapter illustrates two visualization techniques, arrow plots and stream lines, included in the rasterVis package.

11.1 Introduction

The graphics of this chapter display the vector fields defined by the wind direction and speed forecast published in the THREDSS server[1] of Meteogalicia[2]. This server provides access through different protocols to the output of a Weather Research and Forecasting (WRF) model, a mesoscale numerical weather prediction system.

This vector field is encoded in a `RasterStack` object with two layers: wind speed or vector magnitude, and wind direction or vector direction.

```
library(raster)
library(rasterVis)

## Local vector direction
wDir <- raster('data/wDir')/180*pi
## Local vector magnitude
wSpeed <- raster('data/wSpeed')
## Vector field encoded in a RasterStack with two layers,
##    magnitude
## and direction
windField <- stack(wSpeed, wDir)
names(windField) <- c('magnitude', 'direction')
```

11.2 Arrow Plot

A frequent vector visualization technique is the arrow plot, which draws a small arrow at discrete points within the vector field (Figure 11.1). This approach is best suited for small datasets. If the grid of discrete points gets too dense or if the variations in magnitude are too big, the images tend to be visually confusing.

The `rasterVis` package includes the function `vectorplot`, based on `lattice`[3]. This function is able to display vector fields computed from a `Raster` object. Moreover, as the next example illustrates, this function displays `RasterStack` and `RasterBrick` objects with two layers encoding a vector field, slope (local vector magnitude) and aspect (local vector direction).

---

[1]http://mandeo.meteogalicia.es/thredds/catalogos/WRF_2D/catalog.html
[2]http://www.meteogalicia.es
[3]A ggplot2 solution is available in the ggquiver package (https://github.com/mitchelloharawild/ggquiver), although it does not understand Raster objects.

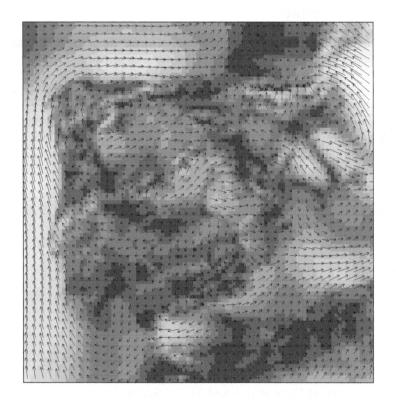

FIGURE 11.1: Arrow plot of the wind vector field.

```
vectorTheme <- BTCTheme(regions = list(alpha = 0.7))

vectorplot(windField,
           isField = TRUE, ##RasterStack is a vector field
           aspX = 5, aspY = 5, ##Multipliers to adjust the relation
                              ##between slope/aspect and
                              ##horizontal/vertical displacements in
                              ##the figure.
           scaleSlope = FALSE, ## Slope values are *not* scaled
           par.settings = vectorTheme,
           colorkey = FALSE,
           scales = list(draw = FALSE))
```

11.3 Streamlines

Another solution is to depict the directional structure of the vector field by its integral curves, also denoted as flow lines or streamlines. There are a variety of algorithms to produce such visualization. The `streamplot` function of `rasterVis` displays streamlines with a procedure inspired by the FROLIC algorithm (Wegenkittl and Gröller 1997): For each point, *droplet*, of a jittered regular grid, a short streamline portion, *streamlet*, is calculated by integrating the underlying vector field at that point. The main color of each streamlet indicates local vector magnitude. Streamlets are composed of points whose sizes, positions, and color degradation encode the local vector direction (Figure 11.2).

```
myTheme <- streamTheme(
    region = rev(brewer.pal(n = 4, "Greys")),
    symbol = rev(brewer.pal(n = 9, "Blues")))

streamplot(windField, isField = TRUE,
        par.settings = myTheme,
        droplet = list(pc = 12), ## Amount of droplets,
            percentage of cells
        streamlet = list(L = 5, ## Length of the streamlet
                    h = 5), ## Calculation step
        scales = list(draw = FALSE),
        panel = panel.levelplot.raster)
```

The magic of Figures 11.1 and 11.2 is that they show the underlying physical structure of the spatial region only displaying wind speed and direction. It is easy to recognize the Iberian Peninsula surrounded by strong winds along the eastern and northern coasts. Another feature easily distinguishable is the Strait of Gibraltar, a channel that connects the Atlantic Ocean to the Mediterranean Sea between the south of Spain and the north of Morocco. Also apparent are the Pyrenees mountains and some of the river valleys.

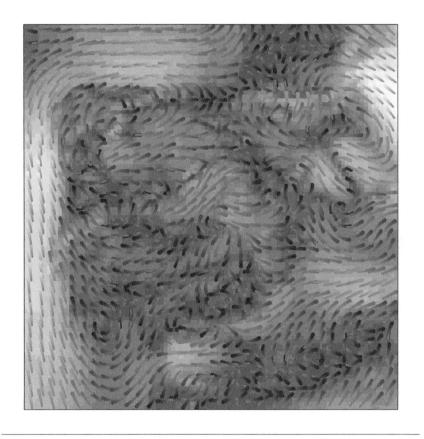

FIGURE 11.2: Streamlines of the wind vector field.

Chapter 12

Physical and Reference Maps

A physical map shows the physical landscape features of a place. Mountains and elevation changes are usually shown with different colors and shades to show relief, using green to show lower elevations and browns for high elevations.

A reference map focuses on the geographic location of features. In these maps, cities are named and major transport routes are identified. In addition, natural features such as rivers and mountains are named, and elevation is shown using a simple color shading.

This chapter details how to create a physical map of Brazil with data from different sources, and a reference map of a northern region of Spain using data from OpenStreetMap.

Next subjects are covered in this chapter: label positioning, hill shading, and map overlaying.

The most relevant packages used in this chapter are: `raster` for reading raster data; `rgdal` for reading shapefiles; `rgeos` for computing intersection between raster and vector data;`rasterVis` for visualization of raster data; `maptools` for label positioning; and `osmdata` for retrieving data from OpenStreetMap.

12.1 Physical Maps

Brazil[1], the world's fifth largest country, is one of the seventeen megadiverse countries[2], home to diverse wildlife, natural environments, and extensive natural resources in a variety of protected habitats. Throughout this section we will create a physical map of this exceptional country using data from several data services.

```
library(raster)
library(rasterVis)
library(rgdal)
library(rgeos)
library(latticeExtra)
library(colorspace)
```

12.1.1 Retrieving Data

Four types of information are needed: administrative boundaries, terrain elevation, rivers and lakes, and sea depth.

1. The administrative boundaries are available from GADM[3]. The readOGR function reads data from the downloaded shapefile and creates a SpatialPolygonsDataFrame object.

```
old <- setwd(tempdir())

download.file('http://biogeo.ucdavis.edu/data/gadm2.8/shp/BRA
    _adm_shp.zip',
             'BRA_adm.zip')
unzip('BRA_adm.zip')
brazilAdm <- readOGR(dsn = '.', layer = 'BRA_adm1')
Encoding(levels(brazilAdm$NAME_1)) <- 'latin1'
```

2. The terrain elevation or digital elevation model (DEM) is available from DIVA-GIS[4]. The raster function reads the file and creates a RasterLayer object.

---

[1]http://en.wikipedia.org/wiki/Brazil
[2]http://en.wikipedia.org/wiki/Megadiverse_countries
[3]http://gadm.org/
[4]http://www.diva-gis.org/Data

```
download.file('http://biogeo.ucdavis.edu/data/diva/alt/BRA_
    alt.zip',
            'BRA_alt.zip')
unzip('BRA_alt.zip')
brazilDEM <- raster('BRA_alt')
```

3. The water lines (rivers and lakes) are available from Natural Earth Data[5]. The readShapeLines function reads data from the downloaded shapefile and creates a SpatialLinesDataFrame object.

```
## World Water lines (Natural Earth)
download.file('http://www.naturalearthdata.com/http//
    www.naturalearthdata.com/download/10m/physical/ne_10m_rivers
    _lake_centerlines.zip',
            'neRivers.zip')
unzip('neRivers.zip')
worldlRiv <- readOGR(dsn = '.', layer = 'ne_10m_rivers_lake_
    centerlines')
```

4. Finally, the sea depth is also available from Natural Earth Data[5]. The raster covers the whole world so it must be cropped by the extent of the DEM raster.

```
download.file('http://www.naturalearthdata.com/http//
    www.naturalearthdata.com/download/10m/raster/OB_LR.zip',
            'neSea.zip')
unzip('neSea.zip')
worldSea <- raster('OB_LR.tif')
brazilSea <- crop(worldSea, brazilDEM)
setwd(old)
```

12.1.2 Intersection of Shapefiles and Elevation Model

The rivers and lakes database from Natural Earth Data comprises all the world extent, but we only need the rivers of Brazil. The function gInter-section of the package rgeos determines the intersection between two geometries. Because these geometries must be defined with classes of the sp package, the extent of brazilDEM must be first converted to Spatial-Polygons. The intersection is a new SpatialLines object, brazilRiv.

---

[5]http://www.naturalearthdata.com/

183

```
## only those features labeled as "River" are needed
worldRiv <- worldRiv[worldRiv$featurecla=='River',]

## Define the extent of Brazil as a SpatialPolygons
extBrazil <- as(extent(brazilDEM), 'SpatialPolygons')
proj4string(extBrazil) <- proj4string(worldRiv)

## and intersect it with worldRiv to extract brazilian rivers
## from the world database
brazilRiv <- gIntersection(worldRiv, extBrazil, byid = TRUE)
## and especially the famous Amazonas River
amazonas <- worldRiv[worldRiv$name=='Amazonas',]
```

12.1.3 Labels

Each region of Brazil will be labeled with the name of its corresponding polygon. The locations of the labels are defined by the centroid of each polygon, easily computed with the `coordinates` method. In addition, a larger label with the name of the country will be placed in the average centroid.

```
## Locations of labels of each polygon
centroids <- coordinates(brazilAdm)
## Location of the "Brazil" label (average of the set of polygons
##     centroids)
xyBrazil <- apply(centroids, 2, mean)
```

Some region names are too long to be displayed in one line. Thus, a previous step is to split the string if it comprises more than two words.

```
admNames <- strsplit(as.character(brazilAdm$NAME_1), ' ')

admNames <- sapply(admNames,
              FUN = function(s){
                  sep = if (length(s)>2) '\n' else ' '
                  paste(s, collapse = sep)
              })
```

12.1.4 Overlaying Layers of Information

Therefore, the physical map (Figure 12.2) is composed of four layers:

1. The sea depth raster displayed with the `levelplot` method of the `rasterVis` package. The palette is defined with `brewer.pal` (Figure 12.1).

```
blueTheme <- rasterTheme(region = brewer.pal(n = 9, 'Blues'))

seaPlot <- levelplot(brazilSea, par.settings = blueTheme,
                maxpixels = 1e6, panel = panel.levelplot.raster,
                margin = FALSE, colorkey = FALSE)
```

1. The altitude raster layer uses a terrain colors palette, as the one produced by the `terrain_hcl` function from the `colorspace` package (Ihaka et al. 2016) (Figure 12.1).

```
terrainTheme <- rasterTheme(region = terrain_hcl(15))

altPlot <- levelplot(brazilDEM, par.settings = terrainTheme,
                maxpixels = 1e6, panel = panel.levelplot.raster,
                margin = FALSE, colorkey = FALSE)
```

2. The rivers represented by the `SpatialLinesDataFrame` object. The Amazonas River is labeled with `sp.lineLabel` and printed with a thicker line. The label is created with the `label` method, a wrapper function to extract the ID slots from the `SpatialLines` and create a suitable character object with the correct names values.

```
amazonasLab <- label(amazonas, 'Amazonas')
```

3. The administrative boundaries represented by the `SpatialPolygons-DataFrame` object with their labels printed with the `panel.pointLabel` function. This function uses optimization routines to find good locations for point labels without overlaps.

```
seaPlot + altPlot + layer({
    ## Rivers
    sp.lines(brazilRiv, col = 'darkblue', lwd = 0.2)
    ## Amazonas
    sp.lineLabel(amazonas, amazonasLab,
            lwd = 1, col = 'darkblue', col.line = 'darkblue',
            cex = 0.5, fontfamily = 'Palatino')
    ## Administrative boundaries
    sp.polygons(brazilAdm, col = 'black', lwd = 0.2)
```

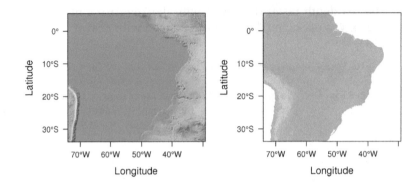

FIGURE 12.1: Sea depth and altitude rasters of Brazil.

```
## Centroids of administrative boundaries ...
panel.points(centroids, col = 'black')
## ... with their labels
panel.pointLabel(centroids, labels = admNames,
         cex = 0.7, fontfamily = 'Palatino', lineheight=
                  .8)
## Country name
panel.text(xyBrazil[1], xyBrazil[2], labels = 'B R A Z I L',
       cex = 1.5, fontfamily = 'Palatino', fontface = 2)
})
```

12.2   Reference maps

Although I was born in Madrid, Galicia (north of Spain) is a very special region for me. More precisely, the Cedeira and Valdoviño regions offer a wonderful combination of wild sea, secluded beaches, and forests. I will show you a map of these marvelous places.

12.2.1   Retrieving Data from OpenStreetMap

The first step is to acquire information from the OpenStreetMap (OSM) project. There are several packages to extract data from this service but, while most of them only provide already rendered raster images, the osm-data package[6] enables the use of the raw data with classes from the packages sp and sf.

---

[6]In the first edition of this book, this chapter was based on the osmar package. However, this package has not been updated since 2013. Moreover, the osmdata package provides an easier query syntax.

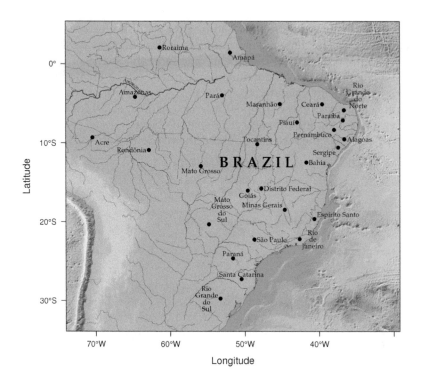

FIGURE 12.2: Physical map of Brazil. Main administrative regions and the Amazonas River are labeled.

osmdata obtains data from the overpass API[7], a read-only API that serves up custom selected parts of the OSM map data. The first step is specifying the bounding box with the function opq:

```
library('osmdata')

## Bounding box
ymax <- 43.7
ymin <- 43.62
xmax <- -8
xmin <- -8.1
## Overpass query
cedeira <- opq(c(xmin, ymin, xmax, ymax))
```

---

[7]http://www.overpass-api.de/

Next, the query is completed adding the required features with add_-osm_feature. The three main arguments of this function are the overpass query defined with opq, the feature key, and the value of this feature. Finally, the information contained in the query can be obtained as a Spatial* object with osmdata_sp, or as a sf object with osmdata_sf. The result is a list with three components, osm_points, osm_lines, and osm_-polygons, containing the respective spatial object.

For example, the next code obtains the residential streets in the region.

```
streetsOSM <- add_osm_feature(cedeira,
                       key = "highway",
                       value = "residential")

streetsSP <- osmdata_sp(streetsOSM)

print(streetsSP)
```

```
Object of class 'osmdata' with:
                 $bbox : 43.6181,-8.0808,43.7031,-8.0224
        $overpass_call : The call submitted to the overpass API
            $timestamp : [ sáb 7 ene 2018 20:02:44 ]
          $osm_points : 'sp' SpatialPointsDataFrame with 819 points
           $osm_lines : 'sp' SpatialLinesDataFrame with 169 lines
        $osm_polygons : 'sp' SpatialPolygonsDataFrame with 3 polygons
       $osm_multilines : 'sp' SpatialNADataFrame with 0 multilines
    $osm_multipolygons : 'sp' SpatialPolygonsDataFrame with 0
                              multipolygons
```

Because this procedure is to be repeated several times, I define a wrapper function that provides a SpatialLinesDataFrame object or a Spatial-PointsDataFrame object depending on the value of its argument type:

```
spFromOSM <- function(source, key, value, type = 'lines')
{
    osm <- add_osm_feature(source, key, value)
    spdata <- osmdata_sp(osm)
    switch(type,
           lines = spdata$osm_lines,
           points = spdata$osm_points)
}
```

The next code uses this function to obtain the different types of roads and streets in the region as SpatialLinesDataFrame objects.

```
streets <- spFromOSM(cedeira, key = "highway", value = "
    residential")
primary <- spFromOSM(cedeira, key = "highway", value = "primary")
secondary <- spFromOSM(cedeira, key = "highway", value = "
    secondary")
tertiary <- spFromOSM(cedeira, key = "highway", value = "tertiary
    ")
unclassified <- spFromOSM(cedeira, key = "highway", value = "
    unclassified")
footway <- spFromOSM(cedeira, key = "highway", value = "footway")
steps <- spFromOSM(cedeira, key = "highway", value = "steps")
```

A similar procedure can be applied to construct a SpatialPointsData-Frame object with the collection of places with name:

```
city <- spFromOSM(cedeira, key = "place", value = "town", type =
    "points")
places <- spFromOSM(cedeira, key = "place", value = "hamlet",
    type = "points")

nms <- strsplit(as.character(places$name), split = ' \\(')
places$name <- sapply(nms, function(x) x[1])
```

12.2.2 Hill Shading

The second step is to produce layers to display the topography. A suitable method is shaded relief or hill shading, previously exposed in section 10.1.1.

The hill shade layer is computed from the slope and aspect layers derived from a Digital Elevation Model. The DEM of Galicia is available at the Geonetwork service of the Xunta de Galicia[8]. I have extracted the data corresponding to the region of interest using crop, and the corresponding files are available at the data folder of the book repository.

```
library(raster)
library(rasterVis)

projCedeira <- projection(city)

demCedeira <- raster('data/demCedeira')
projection(demCedeira) <- projCedeira
```

---

[8]http://xeocatalogo.xunta.es/geonetwork/srv/gl/main.home

```
## Crop the DEM using the bounding box of the OSM data
OSMextent <- extent(extendrange(c(xmin, xmax)),
                    extendrange(c(ymin, ymax)))
demCedeira <- crop(demCedeira, OSMextent)

## Discard values below sea level
demCedeira[demCedeira <= 0] <- NA
```

The slope and aspect layers are computed with the terrain function, and the hill shade layer is derived with these layers for a fixed sun position.

```
slope <- terrain(demCedeira, 'slope')
aspect <- terrain(demCedeira, 'aspect')
hsCedeira <- hillShade(slope = slope, aspect = aspect,
                       angle = 20, direction = 30)
```

12.2.3 Overlaying Layers of Information

And finally, the third step is to display the different layers of information in correct order (Figure 12.3):

- The hill shade layer is created with the levelplot method for Raster objects defined in the rasterVis package. The GrTheme is modified to display the sea region with blue color.

```
## The background color of the panel is set to blue to represent
   the sea
hsTheme <- GrTheme(panel.background = list(col = 'skyblue3'))
```

- The DEM raster is printed with terrain colors and semitransparency over the hill shade layer.

```
library(colorspace)
## DEM with terrain colors and semitransparency
terrainTheme <- rasterTheme(region = terrain_hcl(n = 15),
                            regions = list(alpha = 0.6))
```

- The places are represented with sp.points and labeled with the sp.pointLabel method, a modification of the pointLabel function for base graphics, both defined in the maptools package. These functions use optimization routines to find good locations for point labels without overlaps.

```
library(maptools)
```

- The roads are displayed with an auxiliary function (sp.road) that produces a colored line over a thicker black line.

```
##Auxiliary function to display the roads. A thicker black line
    in
##the background and a thinner one with an appropiate color.
sp.road <- function(line, lwd = 6, blwd = 7,
                col = 'indianred1', bcol = 'black'){
    sp.lines(line, lwd = blwd, col = bcol)
    sp.lines(line, lwd = lwd, col = col)
}


## Hill shade and DEM overlaid
levelplot(hsCedeira, maxpixels = ncell(hsCedeira),
        par.settings = hsTheme,
        margin = FALSE, colorkey = FALSE,
        xlab = '', ylab = '') +
    levelplot(demCedeira, maxpixels = ncell(demCedeira),
            par.settings = terrainTheme) +
    ## Roads and places
    layer({
        ## Street and roads
        sp.road(streets, lwd = 1, blwd = 1, col = 'white')
        sp.road(unclassified, lwd = 2, blwd = 2, col = 'white')
        sp.road(footway, lwd = 2, blwd = 2, col = 'white')
        sp.road(steps, lwd = 2, blwd = 2, col = 'white')
        sp.road(tertiary, lwd = 4, blwd = 4, col = 'palegreen')
        sp.road(secondary, lwd = 6, blwd = 6, col = 'midnightblue')
        sp.road(primary, lwd = 7, blwd = 8, col = 'indianred1')
        ## Places except Cedeira town
        sp.points(places, pch = 19, col = 'black', cex = 0.4, alpha
            = 0.8)
        sp.pointLabel(places, labels = places$name,
```

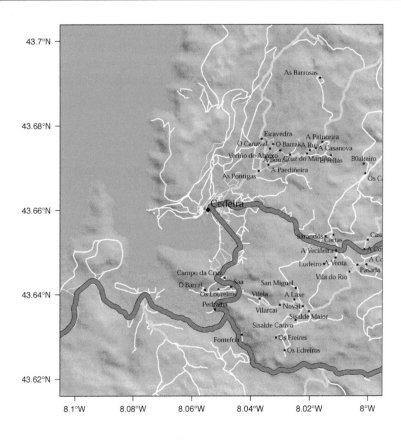

FIGURE 12.3: Main roads near Cedeira, Galicia. Local topography is displayed with the hill shading technique. Some places are highlighted.

```
                    fontfamily = 'Palatino',
                    cex = 0.7, col = 'black')
        ## Cedeira town
        sp.points(city, pch = 18, col = 'black', cex = 1)
        sp.pointLabel(city, labels = 'Cedeira',
                    fontfamily = 'Palatino',
                    cex = 1, col = 'black')
    })
```

Chapter 13

About the Data

13.1 Air Quality in Madrid

Air pollution is harmful to health and contributes to respiratory and cardiac diseases, and has a negative impact on natural ecosystems, agriculture, and the built environment. In Spain, the principal pollutants are particulate matter (PM), tropospheric ozone, nitrogen dioxide, and environmental noise[1].

The surveillance system of the Integrated Air Quality system of the Madrid City Council consists of twenty-four remote stations, equipped with analyzers for gases (NO_X, CO, ozone, BT_X, HCs, SO_2) and particles (PM10, PM2.5), which measure pollution in different areas of the urban environment. In addition, many of the stations also include sensors to provide meteorological data.

The detailed information of each measuring station can be retrieved from its own webpage defined by its station code.

```
## codeStations.csv is extracted from the document
## http://www.mambiente.munimadrid.es/opencms/export/sites/
      default/calaire/Anexos/INTPHORA-DIA.pdf,
## table of page 3.
```

---

[1]http://www.eea.europa.eu/soer/countries/es/

```
codEstaciones <- read.csv2('data/codeStations.csv')
codURL <- as.numeric(substr(codEstaciones$Codigo, 7, 8))

## The information of each measuring station is available at its
##    own webpage, defined by codURL
URLs <- paste('http://www.mambiente.munimadrid.es/opencms/
    opencms/calaire/contenidos/estaciones/estacion', codURL, '
    .html', sep = '')
```

13.1.1 ✒️Data Arrangement

The station webpage includes several tables that can be extracted with the
readHTMLTable function of the XML package. The longitude and latitude
are included in the second table. The ub2dms function cleans this table and
converts the strings to the DMS class defined by the sp package to represent
degrees, minutes, and decimal seconds.

```
library(XML)
library(sp)

## Access each webpage, retrieve tables and extract long/lat
##    data
coords <- lapply(URLs, function(est){
  tables <- readHTMLTable(est)
  location <- tables[[2]]
  ## Clean the table content and convert to dms format
  ub2dms <- function(x){
    ch <- as.character(x)
    ch <- sub(',', '.', ch)
    ch <- sub('O', 'W', ch) ## Some stations use "O" instead of
      "W"
    as.numeric(char2dms(ch, "º", "'", "''"))
  }
  long <- ub2dms(location[2,1])
  lat <- ub2dms(location[2,2])
  alt <- as.numeric(sub(' m.', '', location[2, 3]))

  coords <- data.frame(long = long, lat = lat, alt = alt)

  coords
})

airStations <- cbind(codEstaciones, do.call(rbind, coords))
```

```
## The longitude of "El Pardo" station is wrong (positive
      instead of negative)
airStations$long[22] <- -airStations$long[22]

write.csv2(airStations, file = 'data/airStations.csv')
```

The 2011 air pollution data are available from the Madrid City Council webpage[2] and at the data folder of the book repository. The structure of the file is documented in the INTPHORA-DIA document[3]. The readLines function reads the file and a lapply loop processes each line. The result is stored in the file airQuality.csv

```
rawData <- readLines('data/Datos11.txt')
## This loop reads each line and extracts fields as defined by
    the
## INTPHORA file:
## http://www.mambiente.munimadrid.es/opencms/export/sites/
    default/calaire/Anexos/INTPHORA-DIA.pdf
datos11 <- lapply(rawData, function(x){
    codEst <- substr(x, 1, 8)
    codParam <- substr(x, 9, 10)
    codTec <- substr(x, 11, 12)
    codPeriod <- substr(x, 13, 14)
    month <- substr(x, 17, 18)
    dat <- substr(x, 19, nchar(x))
    ## "N" used for impossible days (31st April)
    idxN <- gregexpr('N', dat)[[1]]
    if (idxN==-1) idxN <- numeric(0)
    nZeroDays <- length(idxN)
    day <- seq(1, 31-nZeroDays)
    ## Substitute V and N with ";" to split data from different
        days
    dat <- gsub('[VN]+', ';', dat)
    dat <- as.numeric(strsplit(dat, ';')[[1]])
    ## Only data from valid days
    dat <- dat[day]
    res <- data.frame(codEst, codParam, ##codTec, codPeriod,
                      month, day, year = 2016,
```

[2]http://www.mambiente.munimadrid.es/opencms/opencms/calaire/consulta/descarga_opendata.html

[3]http://www.mambiente.munimadrid.es/opencms/export/sites/default/calaire/Anexos/INTPHORA-DIA.pdf

```
                    dat)
    })
    datos11 <- do.call(rbind, datos11)

    write.csv2(datos11, 'data/airQuality.csv')
```

13.1.2 Combine Data and Spatial Locations

Our next step is to combine the data and spatial information. The locations are contained in airStations, a data.frame that is converted to an SpatialPointsDataFrame object with the coordinates method.

```
library(sp)

## Spatial location of stations
airStations <- read.csv2('data/airStations.csv')
coordinates(airStations) <- ~ long + lat
## Geographical projection
proj4string(airStations) <- CRS("+proj=longlat +ellps=WGS84 +
    datum=WGS84")
```

On the other hand, the airQuality data.frame comprises the air quality daily measurements. We will retain only the NO_2 time series.

```
## Measurements data
airQuality <- read.csv2('data/airQuality.csv')
## Only interested in NO2
NO2 <- airQuality[airQuality$codParam==8, ]
```

We will represent each station using aggregated values (mean, median, and standard deviation) computed with aggregate:

```
NO2agg <- aggregate(dat ~ codEst, data = NO2,
                FUN = function(x) {
                    c(mean = signif(mean(x), 3),
                      median = median(x),
                      sd = signif(sd(x), 3))
                })
NO2agg <- do.call(cbind, NO2agg)
NO2agg <- as.data.frame(NO2agg)
```

The aggregated values (a data.frame) and the spatial information (a SpatialPointsDataFrame) are combined with the spCbind method from the maptools package to create a new SpatialPointsDataFrame. Previously, the data.frame is reordered by matching against the shared key column (airStations$Codigo and NO2agg$codEst):

```
library(rgdal)
library(maptools)
## Link aggregated data with stations to obtain a
     SpatialPointsDataFrame.
## Codigo and codEst are the stations codes
idxNO2 <- match(airStations$Codigo, NO2agg$codEst)
NO2sp <- spCbind(airStations[, c('Nombre', 'alt')], NO2agg[idxNO2
    , ])
## Save the result
writeOGR(NO2sp, dsn = 'data/', layer = 'NO2sp',
        driver = 'ESRI Shapefile')
```

13.1.3   Photographs of the stations

The photographs of the stations are used for the tooltips of the interactive graphics (Section 8.7). These photographs are downloaded from the Munimadrid webpage[4] with the functions of the XML package.

The htmlParse function from the XML package parses each station page, and the station photograph is extracted with getNodeSet and xmlAttrs.

```
library(XML)

old <- setwd('images')
for (i in 1:nrow(NO2df))
{
    codEst <- NO2df[i, "codEst"]
    ## Webpage of each station
    codURL <- as.numeric(substr(codEst, 7, 8))
    rootURL <- 'http://www.mambiente.munimadrid.es'
    stationURL <- paste(rootURL,
                        '/opencms/opencms/calaire/contenidos/
                           estaciones/estacion',
                        codURL, '.html', sep = '')
    content <- htmlParse(stationURL, encoding = 'utf8')
    ## Extracted with http://www.selectorgadget.com/
    xPath <- '//*[contains(concat( " ", @class, " " ), concat( "
       ", "imagen_1", " " ))]'
    imageStation <- getNodeSet(content, xPath)[[1]]
    imageURL <- xmlAttrs(imageStation)[1]
    imageURL <- paste(rootURL, imageURL, sep = '')
```

---

[4]http://www.mambiente.munimadrid.es/opencms/opencms/calaire/
SistemaIntegral/SistVigilancia/Estaciones/

```
        download.file(imageURL, destfile = paste(codEst, '.jpg', sep =
            ''))
}
setwd(old)
```

13.2   Spanish General Elections

The results from the 2016 Spanish general elections[5] are available from the Ministry webpage[6] and at the data folder of the book repository. Each region of the map will represent the percentage of votes (pcMax) obtained by the predominant political option (whichMax) at the corresponding municipality. Only six groups are considered: the four main parties (PP, PSOE, UP, Cs), the abstention results (ABS), and the remaining parties (OTH). Each region will be identified by the PROVMUN code.

```
dat2016 <- read.csv('data/GeneralSpanishElections2016.csv')

census <- dat2016$Total.censo.electoral
validVotes <- dat2016$Votos.válidos
## Election results per political party and municipality
votesData <- dat2016[, -(1:13)]
## Abstention as an additional party
votesData$ABS <- census - validVotes
## UP is a coalition of several parties
UPcols <- grep("PODEMOS|ECP", names(votesData))
votesData$UP <- rowSums(votesData[, UPcols])
votesData[, UPcols] <- NULL
## Winner party at each municipality
whichMax <- apply(votesData, 1, function(x)names(votesData)[
    which.max(x)])
## Results of the winner party at each municipality
Max <- apply(votesData, 1, max)
## OTH for everything but PP, PSOE, UP, Cs, and ABS
whichMax[!(whichMax %in% c('PP', 'PSOE', 'UP', 'C.s', 'ABS'))] <-
    'OTH'
## Percentage of votes with the electoral census
pcMax <- Max/census * 100

## Province-Municipality code. sprintf formats a number with
    leading zeros.
```

---

[5]https://en.wikipedia.org/wiki/Spanish_general_election,_2016
[6]http://www.infoelectoral.mir.es/infoelectoral/docxl/02_201606_1.zip

```
PROVMUN <- with(dat2016, paste(sprintf('%02d', Có
    digo.de.Provincia),
                        sprintf('%03d', Código.de.Municipio),
                        sep=""))

votes2016 <- data.frame(PROVMUN, whichMax, Max, pcMax)
write.csv(votes2016, 'data/votes2016.csv', row.names = FALSE)
```

```
   PROVMUN      whichMax        Max            pcMax
01001  :  1   ABS :2817   Min.   :     2   Min.   :21.33
01002  :  1   C.s :   3   1st Qu.:    54   1st Qu.:31.69
01003  :  1   OTH :  170  Median :   162   Median :35.64
01004  :  1   PP  :4214   Mean   :  1394   Mean   :37.58
01006  :  1   PSOE: 783   3rd Qu.:   637   3rd Qu.:41.25
01008  :  1   UP  : 138   Max.   :696804   Max.   :94.74
(Other):8119
```

13.2.1   Administrative Boundaries

The Spanish administrative boundaries are available as shapefiles at the INE (Instituto Nacional de Estadística) webpage[7]. Both the municipalities, spMap, and province boundaries, provinces, are read as Spatial-PolygonsDataFrame with readOGR.

```
library(sp)
library(rgdal)

old <- setwd(tempdir())

download.file('ftp://www.ine.es/pcaxis/mapas_completo_
    municipal.rar',
            'mapas_completo_municipal.rar')
system2('unrar', c('e', 'mapas_completo_municipal.rar'))

spMap <- readOGR("esp_muni_0109.shp",
            p4s = "+proj=utm +zone=30 +ellps=GRS80 +units=m +no_
                defs")
Encoding(levels(spMap$NOMBRE)) <- "latin1"

setwd(old)
```

---

[7]http://www.ine.es/ > Products and services > Publications > Download the PC-Axis program > Municipal maps

Some of the polygons are repeated and can be dissolved with union-SpatialPolygons (the rgeos package must be installed).

```
## dissolve repeated polygons
spPols <- unionSpatialPolygons(spMap, spMap$PROVMUN)
```

The main step is to link the data with the polygons. The ID slot of each polygon is the key to find the correspondent registry in the votes2016 dataset.

```
votes2016 <- read.csv('data/votes2016.csv',
                colClasses = c('factor', 'factor', 'numeric',
                'numeric'))

## Match polygons and data using ID slot and PROVMUN column
IDs <- sapply(spPols@polygons, function(x)x@ID)
idx <- match(IDs, votes2016$PROVMUN)

##Places without information
idxNA <- which(is.na(idx))

##Information to be added to the SpatialPolygons object
dat2add <- votes2016[idx, ]

## SpatialPolygonsDataFrame uses row names to match polygons with
    data
row.names(dat2add) <- IDs
spMapVotes <- SpatialPolygonsDataFrame(spPols, dat2add)

## Drop those places without information
spMapVotes0 <- spMapVotes[-idxNA, ]

## Save the result
writeOGR(spMapVotes0, dsn = 'data/', layer = 'spMapVotes0',
    drive = 'ESRI Shapefile')
```

Finally, Spanish maps are commonly displayed with the Canarian islands next to the peninsula. First we have to extract the polygons of the islands and the polygons of the peninsula, and then shift the coordinates of the islands with elide. Finally, a new SpatialPolygons object binds the shifted islands with the peninsula.

```
## Extract Canarias islands from the SpatialPolygons object
canarias <- sapply(spMapVotes0@polygons, function(x)substr(x@ID,
    1, 2) %in% c("35", "38"))
```

```
peninsula <- spMapVotes0[!canarias,]
island <- spMapVotes0[canarias,]

## Shift the island extent box to position them at the bottom
   right corner
dy <- bbox(peninsula)[2,1] - bbox(island)[2,1]
dx <- bbox(peninsula)[1,2] - bbox(island)[1,2]
island2 <- elide(island, shift = c(dx, dy))
bbIslands <- bbox(island2)
proj4string(island2) <- proj4string(peninsula)

## Bind Peninsula (without islands) with shifted islands
spMapVotes <- rbind(peninsula, island2)

## Save the result
writeOGR(spMapVotes, dsn = 'data/', layer = 'spMapVotes',
         drive = 'ESRI Shapefile')
```

13.3 CM SAF

The Satellite Application Facility on Climate Monitoring (CM SAF) is a
joint venture of the Royal Netherlands Meteorological Institute, the Swedish
Meteorological and Hydrological Institute, the Royal Meteorological Institute of Belgium, the Finnish Meteorological Institute, the Deutscher Wet-
terdienst, Meteoswiss, and the UK MetOffice, along with collaboration of
the European Organization for the Exploitation of Meteorological Satellites (EUMETSAT) (CM SAF 2013). The CM-SAF was funded in 1992 to
generate and store monthly and daily averages of meteorological data
measured in a continuous way with a spatial resolution of 0.03° (15 kilometers). The CM SAF provides two categories of data: operational products and climate data. The operational products are built on data that are
validated with on-ground stations and then is provided in near-real-time
to develop variability studies in diurnal and seasonal time scales. However, climate data are long-term data series to assess inter-annual variability (Posselt, Mueller, et al. 2012).

In this chapter we will display the annual average of the shortwave
incoming solar radiation product (SIS) incident over Spain during 2008,
computed from the monthly means of this variable. SIS collates shortwave
radiation (0.2 to 4 µm wavelength range) reaching a horizontal unit Earth
surface obtained by processing information from geostationary satellites

(METEOSAT) and also from polar satellites (MetOp and NOAA) (Schulz et al. 2009) and then validated with high-quality on-ground measurements from the Baseline Surface Radiation Network (BSRN)[8].

The monthly means of SIS are available upon request from the CM SAF webpage (Posselt, Müller, et al. 2011) and at the data folder of the book repository. Data from CM-SAF is published as raster files using the NetCDF format. The raster package provides the stack function to read a set of files and create a RasterStack object, where each layer stores the content of a file. Therefore, the twelve raster files of monthly averages produce a RasterStack with twelve layers.

```
library(raster)

tmp <- tempdir()
unzip('data/SISmm2008_CMSAF.zip', exdir = tmp)
filesCMSAF <- dir(tmp, pattern = 'SISmm')
SISmm <- stack(paste(tmp, filesCMSAF, sep = '/'))
## CM-SAF data is average daily irradiance (W/m2). Multiply by
     24
## hours to obtain daily irradiation (Wh/m2)
SISmm <- SISmm * 24
```

The RasterLayer object with annual averages is computed from the monthly means and stored using the native format of the raster package.

```
## Monthly irradiation: each month by the corresponding number
     of days
daysMonth <- c(31, 29, 31, 30, 31, 30, 31, 31, 30, 31, 30, 31)
SISm <- SISmm * daysMonth / 1000 ## kWh/m2
## Annual average
SISav <- sum(SISm)/sum(daysMonth)
writeRaster(SISav, file = 'SISav')
```

13.4   Land Cover and Population Rasters

The NASA's Earth Observing System (EOS)[9] is a coordinated series of polar-orbiting and low-inclination satellites for long-term global observations of the land surface, biosphere, solid Earth, atmosphere, and oceans. NEO-NASA[10], one of projects included in EOS, provides a repository of

---

[8]http://www.bsrn.awi.de/en/home/
[9]http://eospso.gsfc.nasa.gov/
[10]http://neo.sci.gsfc.nasa.gov

global data imagery. We use the population density and land cover classification rasters. Both rasters must be downloaded from their respective webpages as Geo-TIFF files.

```
library(raster)
## http://neo.sci.gsfc.nasa.gov/Search.html?group=64
pop <- raster('875430rgb-167772161.0.FLOAT.TIFF')
## http://neo.sci.gsfc.nasa.gov/Search.html?group=20
landClass <- raster('241243rgb-167772161.0.TIFF')
```

Part III

Space-Time Data

Chapter 14

Displaying Spatiotemporal Data: Introduction

Space-time datasets are indexed in both space and time. The data may consist of a spatial vector object (for example, points or polygons) or raster data at different times. The first case is representative of data from fixed sensors providing measurements abundant in time but sparse in space. The second case is the typical format of satellite imagery, which produces high spatial resolution data sparse in time (E. Pebesma 2012).

There are several visualization approaches of space-time data trying to cope with the four dimensions of the data (Cressie and C. Wikle 2015).

On the one hand, the data can be conceived as a collection of snapshots at different times. These snapshots can be displayed as a sequence of frames to produce an animation, or can be printed on one page with different panels for each snapshot using the small-multiple technique described repeatedly in previous chapters.

On the other hand, one of the two spatial dimensions can be collapsed through an appropriate statistic (for example, mean or standard deviation) to produce a space-time plot (also known as a Hovmöller diagram). The axes of this graphic are typically longitude or latitude as the x-axis, and time as the y-axis, with the value of the spatial-averaged value of the raster data represented with color.

Finally, the space-time object can be reduced to a multivariate time series (where each location is a variable or column of the time series) and displayed with the time series visualization techniques described in the Part I. This approach is directly applicable to space-time data sparse in space (for example, point measurements at different times). However, it is mandatory to use aggregation in the case of raster data. In this case, the multivariate time series is composed of the evolution of the raster data averaged along a certain direction.

The next chapters, focused on raster space-time data (Chapters 15 and 17) and point space-time data (Chapter 16), illustrate with examples how to produce animations, multipanel graphics, hovmöller diagrams, and time-series with R.

14.1 Packages

The CRAN Tasks View "Handling and Analyzing Spatiotemporal Data" [1] summarizes the packages for reading, vizualizing, and analyzing space-time data. This section provides a brief introduction to the spacetime, raster, and rasterVis packages. Most of the information has been extracted from their vignettes, webpages, and help pages. You should read them for detailed information.

14.1.1 spacetime

The spacetime package (E. Pebesma 2012) is built upon the classes and methods for spatial data from the sp package , and for time series data from the xts package. It defines classes to represent four space-time layouts:

1. STF, STFDF: full space-time grid of observations for spatial features and observation time, with all space-time combinations.

2. STS, STSDF: sparse grid layout, stores only the non-missing space-time combinations on a lattice

3. STI, STIDF: irregular layout, time and space points of measured values have no apparent organisation.

4. STT, STTDF: simple trajectories.

---

[1]http://cran.r-project.org/web/views/SpatioTemporal.html

Moreover, spacetime provides several methods for the following classes:

- stConstruct, STFDF, and STIDF create objects from single or multiple tables.

- as coerces to other spatiotemporal objects, xts, Spatial, matrix, or data.frame.

- [[selects or replaces data values.

- [selects spatial or temporal subsets, and data variables.

- over retrieves index or data values of one object at the locations and times of another.

- aggregate aggregates data values over particular spatial, temporal, or spatiotemporal domains.

- stplot creates spatiotemporal plots. It is able to produce multipanel plots, space-time plots, animations, and time series plots.

14.1.2 raster

The raster package (R. J. Hijmans 2017) is able to add time information associated with layers of a RasterStack or RasterBrick object with the setZ function. This information can be extracted with getZ.

If a Raster* object includes this information, the zApply function can be used to apply a function over a time series of layers of the object.

14.1.3 rasterVis

rasterVis (Perpiñán and R. Hijmans 2017) provides three methods to display spatiotemporal rasters:

1. hovmoller produces Hovmöller diagrams (Hovmöller 1949). The axes of this kind of diagram are typically longitude or latitude (x-axis) and time (ordinate or y-axis) with the value of some aggregated field represented through color. However, the user can define the direction with dirXY and the summary function with FUN.

2. horizonplot creates horizon graphs (Few 2008), with many time series displayed in parallel by cutting the vertical range into segments and overplotting them with color representing the magnitude and direction of deviation. Each time series corresponds to a geographical zone defined with dirXY and averaged with zonal.

3. xyplot displays conventional time series plots. Each time series corresponds to a geographical zone defined with dirXY and aggregated with zonal.

On the other hand, the histogram, densityplot, and bwplot methods accept a FUN argument to be applied to the z slot of Raster* object (defined by setZ). The result of this function is used as the grouping variable of the plot to create different panels.

14.1.4   rgl

rgl is a package that produces real-time interactive 3D plots. It allows to interactively rotate, zoom the graphics and select regions. This package uses the OpenGL[2] library as the rendering backend providing an interface to graphics hardware. It contains high-level graphics functions similar to base R graphics, but working in three dimensions. Moreover, it provides low level functions inspired by the grid package.

14.2   Further Reading

- (Cressie and C. Wikle 2015) is a systematic approach to key quantitative techniques on statistics for spatiotemporal data. The book begins with separate treatments of temporal data and spatial data, and later combines these concepts to discuss spatiotemporal statistical methods. There is a chapter devoted to exploratory methods, including visualization techniques.

- (E. Pebesma 2012) presents the spacetime package, which implements a set of classes for spatiotemporal data. This paper includes examples that illustrate how to import, subset, coerce, and export spatiotemporal data, proposes several visualization methods, and discusses spatiotemporal geostatistical interpolation.

- (Slocum 2005) (previously cited in Chapter 7.2) includes a chapter about map animation, discussing several approaches for displaying spatiotemporal data.

- (Hengl 2009) (previously cited in Chapter 7.2) includes a working example with spatiotemporal data to illustrate space-time variograms and interpolation.

---

[2]https://www.opengl.org/

- (Harrower and Fabrikant 2008) explore the role of animation in geographic visualization and outline the challenges, both conceptual and technical, involved in the creation and use of animated maps.

- The CRAN Tasks View "Handling and Analyzing Spatiotemporal Data" [3] summarizes the packages for reading, vizualizing, and analyzing space-time data. The R-SIG-Geo mailing list [4] is a powerful resource for obtaining help.

---

[3] http://cran.r-project.org/web/views/SpatioTemporal.html
[4] https://stat.ethz.ch/mailman/listinfo/R-SIG-Geo/

Chapter 15

Spatiotemporal Raster Data

A space-time raster dataset is a collection of raster layers indexed by time, or in other words, a time series of raster maps. The raster package defines the classes RasterStack and RasterBrick to build multilayer rasters. The index of the collection can be set with the function setZ (which is not restricted to time indexes). The raster-Vis package provide several methods to display space-time rasters. This chapter covers these subjects: small multiples; scatterplot matrices; horizon graph; and interactive and 3D visualization.

The most relevant packages in this chapter are: raster for reading raster data; zoo for definition of time series; rasterVis for displaying raster data; mapview for interactive graphics.

15.1 Introduction

Throughout this chapter we will work with a multilayer raster of daily solar radiation estimates from CM SAF (section 13.3) falling in the region of Galicia (north of Spain) during 2011. These data are arranged in a Raster-Brick with 365 layers using brick and time indexed with setZ.

```
library(raster)
library(zoo)
library(rasterVis)

SISdm <- brick('data/SISgal')

timeIndex <- seq(as.Date('2011-01-01'), by = 'day', length = 365)
SISdm <- setZ(SISdm, timeIndex)
names(SISdm) <- format(timeIndex, '%a_%Y%m%d')

class : RasterBrick
dimensions : 70, 95, 6650, 365 (nrow, ncol, ncell, nlayers)
resolution : 0.03, 0.03 (x, y)
extent : -9.385, -6.535, 41.735, 43.835 (xmin, xmax, ymin, ymax)
coord. ref. : +proj=longlat +datum=WGS84 +ellps=WGS84 +towgs84
    =0,0,0
data source : data/SISgal.grd
names : sáb_20110101, dom_20110102, lun_20110103, mar_20110104,
    mié_20110105, jue_20110106, vie_20110107, sáb_20110108, dom_
    20110109, lun_20110110, mar_20110111, mié_20110112, jue_
    20110113, vie_20110114, sáb_20110115, ...
min values : 22.368380, 17.167900, 14.132746, 6.816897,
    11.616024, 4.851091, 7.251504, 7.717323, 32.492584,
    16.786699, 20.433887, 25.984509, 24.190809, 25.485737,
    31.256468, ...
max values : 88.84628, 93.43065, 84.44052, 81.79698, 47.08540,
    49.01350, 80.98129, 74.12977, 102.71733, 48.54120, 75.46757,
    98.09704, 104.65322, 104.29843, 107.23942, ...
time : 2011-01-01, 2011-12-31 (min, max)
```

15.2 Level Plots

This multilayer raster can be displayed with each snapshot in a panel using the small-multiple technique. The problem with this approach is that only a limited number of panels can be correctly displayed on one page. In this example, we print the first 12 days of the sequence (Figure 15.1).

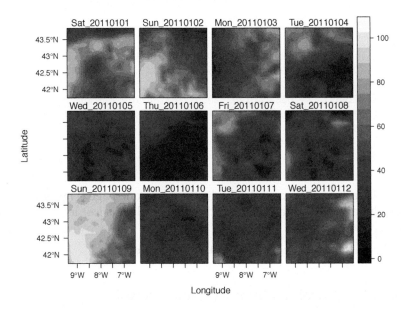

FIGURE 15.1: Level plot of daily averages of solar radiation.

```
levelplot(SISdm, layers = 1:12, panel = panel.levelplot.raster)
```

When the number of layers is very high, a partial solution is to aggregate the data, grouping the layers according to a time condition. For example, we can build a new space-time raster with the monthly averages using zApply and as.yearmon.

```
SISmm <- zApply(SISdm, by = as.yearmon, fun = 'mean')

class : RasterBrick
dimensions : 70, 95, 6650, 12 (nrow, ncol, ncell, nlayers)
resolution : 0.03, 0.03 (x, y)
extent : -9.385, -6.535, 41.735, 43.835 (xmin, xmax, ymin, ymax)
```

```
coord. ref. : +proj=longlat +datum=WGS84 +ellps=WGS84 +towgs84
    =0,0,0
data source : in memory
names : ene.2011, feb.2011, mar.2011, abr.2011, may.2011, jun
    .2011, jul.2011, ago.2011, sep.2011, oct.2011, nov.2011, dic
    .2011
min values : 45.79540, 90.73799, 127.99301, 206.37934, 171.51269,
    233.96372, 196.51799, 188.50950, 161.11302, 126.57236,
    64.57646, 49.24799
max values : 74.62326, 123.82005, 167.88226, 249.85596,
    313.51773, 340.51255, 334.61417, 273.41496, 225.10363,
    164.54701, 82.47245, 79.40128
        : ene 2011, feb 2011, mar 2011, abr 2011, may 2011, jun
            2011, jul 2011, ago 2011, sep 2011, oct 2011, nov
            2011, dic 2011
```

This raster can be completely displayed on one page (Figure 15.2), although part of the information of the original data is lost in the aggregation procedure.

```
levelplot(SISmm, panel = panel.levelplot.raster)
```

15.3 Graphical Exploratory Data Analysis

There are other graphical tools that complement the previous maps. The scatterplot and the matrix of scatterplots, the histogram and kernel density plot, and the boxplot are among the most important tools in the frame of the Exploratory Data Analysis approach. Some of them were previously used with a spatial raster (Chapter 10). In this section we will use the histogram (Figure 15.3),

```
histogram(SISdm, FUN = as.yearmon)
```

the violin plot (a combination of a boxplot and a kernel density plot) (Figure 15.4),

```
bwplot(SISdm, FUN = as.yearmon)
```

and the matrix of scatterplots (section 4.1, Figure 15.5).

```
splom(SISmm, xlab = '', plot.loess = TRUE)
```

Both the histogram and the violin plot show that daily solar irradiation is bimodal almost every month. This is related to the predominance of clear sky and overcast days, with several partly cloudy days between these

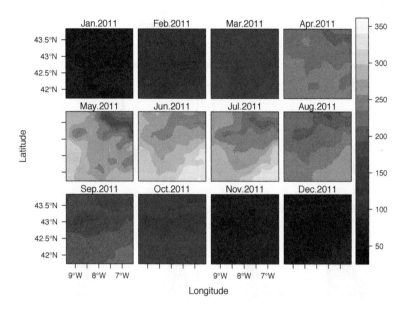

FIGURE 15.2: Level plot of monthly averages of solar radiation.

modes. This geographical region receives higher irradiation levels from June to September, and both the levels and the shape of the probability distribution contrast sharply with the winter.

The matrix of scatterplots displays a quasilinear relationship between the summer months due to the predominance of clear sky conditions. However, the relationships involving winter months become strongly non-linear due to the presence of clouds.

15.4 Space-Time and Time Series Plots

The level plots of Figures 15.1 and 15.2 display the full 3D space-time data using a grid of panels where each layer is depicted in a separate panel. In

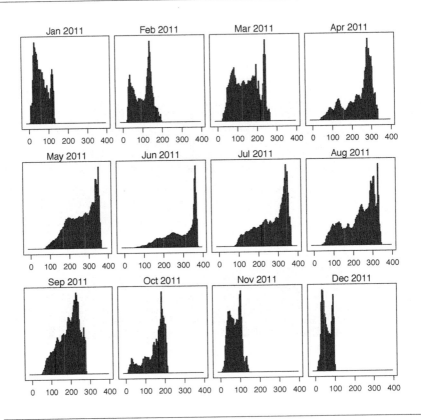

Figure 15.3: Histogram of monthly distribution of solar radiation.

the section 17.1, this collection of layers will be displayed sequentially like frames of a movie to build an animation. In this section, the 3D raster is reduced to a 2D matrix with spatial aggregation following a certain direction. For example, Figure 15.6 displays with colors the averaged value of the raster for each latitude zone (using the default value of the argument dirXY) with time on the vertical axis.

```
hovmoller(SISdm)
```

On the other hand, this 2D matrix can be conceived as a multivariate time series with each aggregated zone conforming to a different variable of the time series. This approach is followed by the xyplot (Figure 15.7) and horizonplot (Figure 15.8) methods, which reproduce the procedures described in Chapter 3 to display multivariate time series.

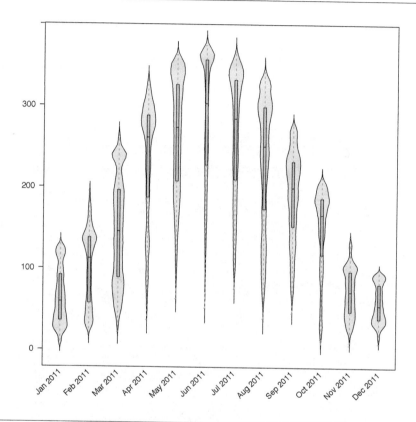

FIGURE 15.4: Violin plot of monthly distribution of solar radiation.

```
xyplot(SISdm, auto.key = list(space = 'right'))

horizonplot(SISdm, digits = 1,
        col.regions = rev(brewer.pal(n = 6, 'PuOr')),
        xlab = '', ylab = 'Latitude')
```

These three figures highlight the stational behavior of the solar radiation, with higher values during the summer. It is interesting to note that (Figure 15.8) the radiation values around the equinoxes fluctuate near the yearly average value of each latitude region.

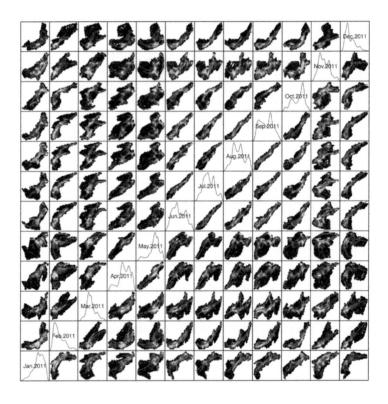

FIGURE 15.5: Scatterplot matrix of monthly averages together with their kernel density estimations in the diagonal frames.

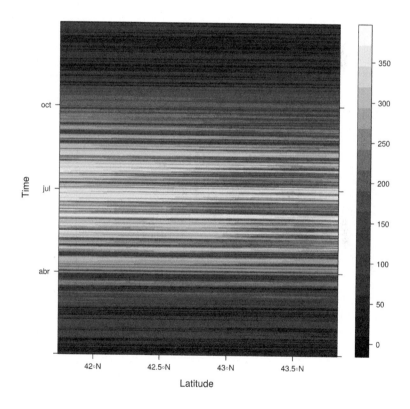

FIGURE 15.6: Hovmöller graphic displaying the time evolution of the average solar radiation for each latitude zone.

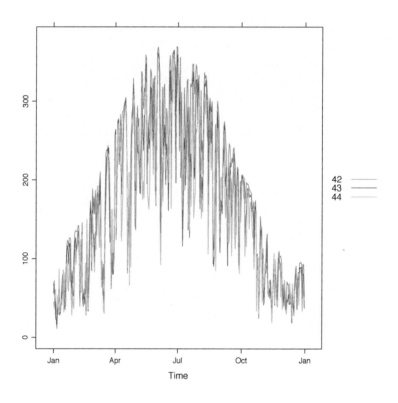

FIGURE 15.7: Time graph of the average solar radiation for each latitude zone. Each line represents a latitude band.

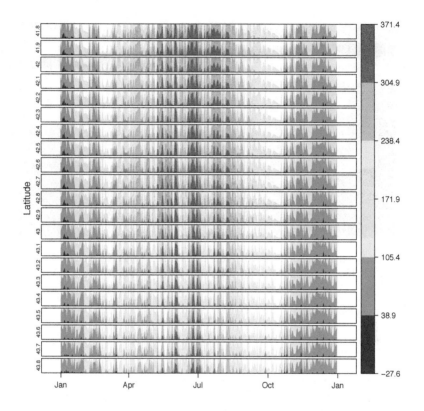

FIGURE 15.8: Horizon graph of the average solar radiation for each latitude zone.

15.4.1 Interactive graphics: cubeView

Figure 15.6 reduces the 3D raster to a 2D matrix with spatial aggregation following a certain direction. The `mapview` package provides a function, `cubeView`, able to represent this 3D raster without prior aggregation, as an interactive cube.

This cube can be freely rotated so that different Hövmoller views are possible. Visible layers can be selected using arrow keys (left-right for x-axis, up-down for y-axis), and PageUp-PageDown keys for z-axis. Using the mouse, the cube can be rotated with left button, moved with the right button, and zoom using the mouse wheel. Figure 15.9 shows a snapshot of the cube produced with the next code.

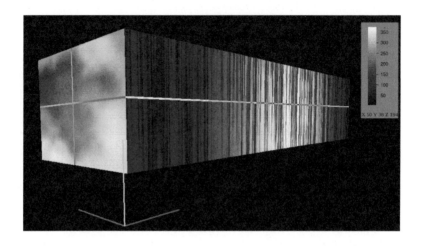

FIGURE 15.9: Snapshot of an interactive cube displaying a 3D raster.

```
library(mapview)

cubeView(SISdm)
```

Chapter 16

Spatiotemporal Point Observations

Throughout this chapter we will revisit the data from the Integrated Air Quality system of the Madrid City Council (Section 13.1) to illustrate visualization methods applicable for point space-time data. This dataset comprises the time series of measurements acquired at each station of the network during 2011. In Section 8 the data were converted from spatiotemporal data to spatial data, where the time information was suppressed to display only the yearly average values. In this chapter we will work with the whole space-time dataset using the tools provided by the spacetime package (E. Pebesma 2012).

Most relevant packages in this chapter are: sp, zoo, and spacetime for reading and defining point space-time data; reshape2 for converting from long to wide format.

16.1 Introduction

The starting point is to retrieve the data and combine it with the spatial and temporal information. The data are contained in the airQuality data.frame, and the locations are in airStations, a data.frame that is converted to a SpatialPointsDataFrame object with the coordinates method.

```
library(sp)

## Spatial location of stations
airStations <- read.csv2('data/airStations.csv')
## rownames are used as the ID of the Spatial object
rownames(airStations) <- substring(airStations$Codigo, 7)
coordinates(airStations) <- ~ long + lat
proj4string(airStations) <- CRS("+proj=longlat +ellps=WGS84")
## Measurements data
airQuality <- read.csv2('data/airQuality.csv')
## Only interested in NO2
NO2 <- airQuality[airQuality$codParam == 8, ]
```

Each row of the NO2 data.frame corresponds to a measurement at one of the stations during a day of the year (long format, following the schema proposed in (E. Pebesma 2012)).

The spacetime package defines several classes for spatiotemporal data inheriting the classes defined by the sp and xts packages. In particular, the STFDF, a class for spatiotemporal data with full space-time grids with n spatial locations and m times, requires a data.frame with $n \cdot m$ rows, (spatial index moving faster than temporal index). Thus, we need to transform this structure to build a multivariate time series where each station is a different variable (space-wide under the schema of (E. Pebesma 2012)). The procedure is

- Add a column with the POSIXct time index (line 5)

- Reshape the data.frame from long to wide format with dcast (line 8).

- Define a multivariate time series with zoo (Figure 16.3, line 12)

- Coerce this time series to a vector with $n \cdot m$ rows (line 14).

- Finally, create the STFDF object with the previous components (line 16).

```
 1  library(zoo)
 2  library(reshape2)
 3  library(spacetime)
 4
 5  NO2$time <- as.Date(with(NO2,
 6                      ISOdate(year, month, day)))
 7
 8  NO2wide <- dcast(NO2[, c('codEst', 'dat', 'time')],
 9              time ~ codEst,
10              value.var = "dat")
11
12  NO2zoo <- zoo(NO2wide[,-1], NO2wide$time)
13
14  dats <- data.frame(vals = as.vector(t(NO2zoo)))
15
16  NO2st <- STFDF(sp = airStations,
17              time = index(NO2zoo),
18              data = dats)
```

16.2 Graphics with `spacetime`

The `stplot` function of the `spacetime` package supplies the main visualization methods for spatiotemporal data. When the mode xy is chosen (default) it is mainly a wrapper around `spplot` and displays a panel with the spatial data for each element of the time index (Figure 16.1). The problem with this approach is that only a limited number of panels can be correctly displayed on one page. In this example, we print the first twelve days of the sequence.

```
airPal <- colorRampPalette(c('springgreen1', 'sienna3', 'gray5'))
     (5)

stplot(NO2st[, 1:12],
     cuts = 5,
     col.regions = airPal,
     main = '',
     edge.col = 'black')
```

With the mode xt, a space-time plot with space on the x-axis and time on the y-axis is plotted (Figure 16.2).

```
stplot(NO2st, mode = 'xt',
     col.regions = colorRampPalette(airPal)(15),
```

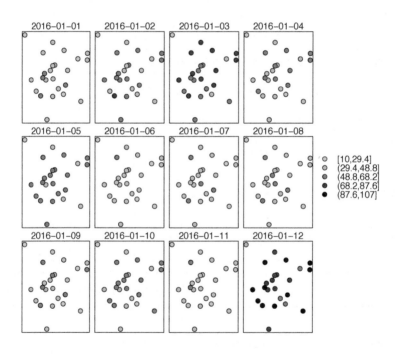

FIGURE 16.1: Scatterplots of the NO_2 values (2011) with a panel for each day of the time series. Each circle represents a different station.

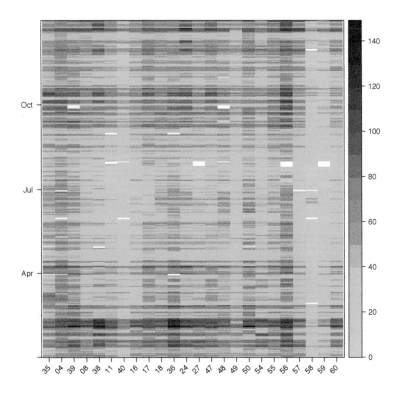

FIGURE 16.2: Space-time graphic of the NO_2 time series. Each column represents a different station (denoted with the last two digits of the code).

```
        scales = list(x = list(rot = 45)),
        ylab = '', xlab = '', main = '')
```

Finally, with the mode `ts`, data are coerced to a multivariate time series that is displayed in a single plot (Figure 16.3).

```
stplot(NO2st, mode = 'ts',
       xlab = '',
       lwd = 0.1, col = 'black', alpha = 0.6,
       auto.key = FALSE)
```

These three graphics complement each other and together provide a more complete view of the behavior of the data. For example in Figure

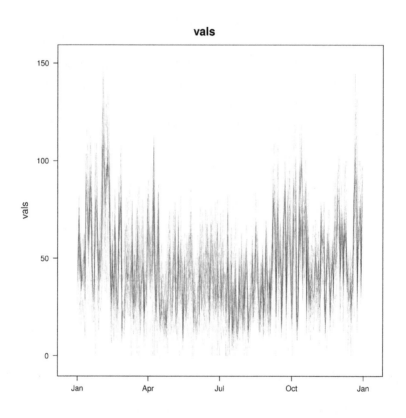

FIGURE 16.3: Time graph of the NO_2 time series (2011). Each line represents a different station.

16.1, we can find stations whose levels remain almost constant through-out the twelve days period (namely, El Pardo-28079058[1], the station at the top-left corner that is far from the city center), while others fluctuate no-tably during this same period (for example, Barajas-28079027 and Urb. Embajada-28079055, the two nearby stations at the right). On the other hand, Figure 16.2 loses the spatial information but gives a more compre-hensive view of the evolution of the network of stations. The station El Pardo-28079058 is significantly below the rest of the stations during the whole year, with the station Pza. Fdez Ladreda-28079056 being the oppo-site. In between, the stations could be divided into two or three groups according to their levels. Regardless, the network of stations reaches max-imum values during the first days of autumn and at the end of winter. These maxima are clearly displayed in Figure 16.3.

---

[1]Use Figure 8.5 as reference of the positions and codes of the stations.

Chapter 17

Animation

In the Chapter 15 a multilayer raster is displayed as a 3D object with a matrix of levelplots using the small-multiple technique (Section 15.2) and as a 2D object after aggregation following a certain direction (Section 15.4). This chapter displays a multilayer raster with a different approach, plotting the individual layers of the space-time raster sequentially as movie frames to produce an animation (Section 17.1).

On the other hand, in Chapter 16 point space-time data is represented with small multiples, space-time graphics and time graphs. In this chapter the time evolution of the data is represented with dynamic bubbles whose radius and color vary over time.

Notwithstanding, the animation technique is not restricted to displaying the evolution of a variable over time. Animations can also be constructed from a sequence of changes in an attribute of the variable: for example, the user's view of an static variable can be gradually moved to create a fly-by animation (Section 17.3).

The most relevant packages in this chapter are: raster for reading raster data; rasterVis for visualization of raster data; maps, mapdata, and maptools for boundaries lines; rgl for 3D visualization; and gridSVG for interactive graphics.

17.1 Depicting Variable Changes Over Time: Raster Data

This section uses animation to visualize changes of a raster variable over time. The procedure is quite simple:

- Plot each layer of the raster to produce a collection of graphic files.

- Join these files as a sequence of frames with a suitable tool (for example, ffmpeg[1]) to create a movie file[2, 3].

The effectiveness of this visualization procedure is partly related to the similitude between consecutive frames. If the frames of the sequence diverge excessively from one to another, the user will experience difficulties to perceive any relationship between them. On the other hand, if the transitions between layers are smooth enough, the frames will be perceived as conforming to a whole story; and, moreover, the user will be able to spot both the stable patterns and the important variations.

17.1.1 Data

The daily solar radiation CM-SAF data (Chapter 15) do not meet the condition of a smooth transition between layers. The changes between the consecutive snapshots of daily radiation are too abrupt to be glued one after another. We will work with a different dataset in this section.

The THREDSS server[4] of Meteogalicia[5] provides access through different protocols to the output of a Weather Research and Forecasting (WRF) model, a mesoscale numerical weather prediction system. Among the set of available variables we will use the forecast of hourly cloud cover at low and mid levels. This space-time raster has a time horizon of 96 hours and a spatial resolution of 12 kilometers.

```
library(raster)
library(rasterVis)

cft <- brick('data/cft_20130417_0000.nc')
```

---

[1]http://www.ffmpeg.org/

[2]The animation package (Xie 2013) defines several functions to wrap ffmpeg and convert from ImageMagick.

[3]An alternative method is the LaTeX animate package, which provides an interface to create portable JavaScript-driven PDF animations from rasterized image files.

[4]http://mandeo.meteogalicia.es/thredds/catalogos/WRF_2D/catalog.html

[5]http://www.meteogalicia.es

```
## set projection
projLCC2d <- "+proj=lcc +lon_0=-14.1 +lat_0=34.823 +lat_1=43 +lat
    _2=43 +x_0=536402.3 +y_0=-18558.61 +units=km +ellps=WGS84"
projection(cft) <- projLCC2d
##set time index
timeIndex <- seq(as.POSIXct('2013-04-17 01:00:00', tz = 'UTC'),
    length = 96, by = 'hour')
cft <- setZ(cft, timeIndex)
names(cft) <- format(timeIndex, 'D%d_H%H')
```

17.1.2 Spatial Context: Administrative Boundaries

Let's provide the spatial context with the countries boundaries, extracted from the worldHires database of the maps and mapdata packages.

```
library(maptools)
library(rgdal)
library(maps)
library(mapdata)

## Project the extent of the cft raster to longitude-latitude,
    because
## the map package works with it.
projLL <- CRS('+proj=longlat +datum=WGS84 +ellps=WGS84 +towgs84
    =0,0,0')
cftLL <- projectExtent(cft, projLL)
cftExt <- as.vector(bbox(cftLL))
## Extract the lines from the map package using this extent
boundaries <- map('worldHires',
                xlim = cftExt[c(1, 3)], ylim = cftExt[c(2, 4)],
                plot = FALSE)
## Convert the result to a SpatialLines object
boundaries <- map2SpatialLines(boundaries, proj4string = projLL)
## Project to the projection of the cft object
boundaries <- spTransform(boundaries, CRS(projLCC2d))
```

17.1.3 Producing the Frames and the Movie

The next step is to produce the collection of frames. We will create a file with each layer of the RasterBrick using the levelplot function. This function provides the argument layout to control the arrangement of a multipanel display. If it is set to c(1,1), a different page is created for each layer.

```
cloudTheme <- rasterTheme(region = brewer.pal(n = 9, 'Blues'))

tmp <- tempdir()
trellis.device(png, file = paste0(tmp, '/Rplot%02d.pdf'),
          res = 300, width = 1500, height = 1500)
levelplot(cft, layout = c(1, 1), par.settings = cloudTheme) +
    layer(sp.lines(boundaries, lwd = 0.6))
dev.off()
```

A suitable tool to concatenate these frames and create the movie is ffm-peg, a free cross-platform software to record, convert, and stream audio and video[1]. The resulting movie is available from the book website.

```
old <- setwd(tmp)
## Create a movie with ffmpeg ...
system2('ffmpeg',
        c('-r 6', ## with 6 frames per second
          '-i Rplot%02d.pdf', ## using the previous files
          '-b:v 300k', ## with a bitrate of 300kbs
          'output.mp4')
        )
file.remove(dir(pattern = 'Rplot'))
file.copy('output.mp4', paste0(old, '/figs/cft.mp4'), overwrite =
    TRUE)
setwd(old)
```

17.1.4   Static Image

Figure 17.1 shows a sequence of twenty-four snapshots (second day of the forecast series) of the movie. This graphic is also created with levelplot but now using the argument layers to choose a subset of the layers, and with a different value for layout to display a matrix of twenty-four panels.

```
levelplot(cft,
        layers = 25:48, ## Layers to display (second day)
        layout = c(6, 4), ## Layout of 6 columns and 4 rows
        par.settings = cloudTheme,
        names.attr = paste0(sprintf('%02d', 1:24), 'h'),
        panel = panel.levelplot.raster) +
    layer(sp.lines(boundaries, lwd = 0.6))
```

The movie and the static image are complementary tools and should be used together. Watching the movie you will perceive the cloud transit

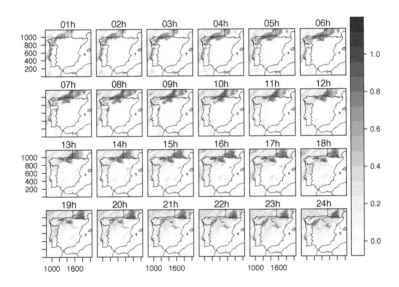

FIGURE 17.1: Forecast of hourly cloud cover at low and mid levels.

from Galicia to the Pyrenees gradually dissolving over the Cantabrian region. On the other hand, with Figure 17.1 you can locate the position of a group of clouds in a certain hour and simultaneously observe the relationship of that position with the evolution during that period. With the movie you will concentrate your attention on the movement. With small multiple pictures, your focus will be on positions and relations. You should use both graphical tools to grasp the entire 3D dataset.

17.1.5   3D animation

In section 10.4.1 an interactive 3D plot of a Digital Elevation Model was produced with the `rgl` package, a visualization device system for R us-

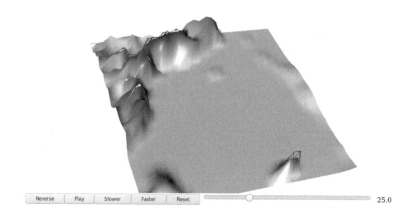

FIGURE 17.2: 3D animation of the forecast of hourly cloud cover at low and mid levels.

ing OpenGL as the rendering backend. With the next code this package generates a 3D animation depicting the cloud evolution over time.

```
library(rgl)

clear3d()

pal <- colorRampPalette(brewer.pal(n = 9, 'Blues'))

N <- nlayers(cft)

ids <- lapply(seq_len(N),
        FUN = function(i)
            plot3D(cft[[i]],
                maxpixels = 1e3,
                col = pal,
                adjust = FALSE, ## Disable automatic scaling
                    of xy axes.
                zfac = 200)) ## Common z scale for all
                graphics

rglwidget() %>%
    playwidget(start = 0, stop = N,
            subsetControl(1, subsets = ids))
```

17.2   ✦Depicting Variable Changes Over Time: Point Space-Time Data

The procedure for point space-time data is more complex than for raster data. This section details a method built over the functionalities of the gridSVG package.

17.2.1   Initial Snapshot

The first step is to define the initial parameters of the animation: starting values and duration.

```
library(gridSVG)
## Initial parameters
start <- NO2st[,1]
## values will be encoded as size of circles,
## so we need to scale them
startVals <- start$vals/5000

nStations <- nrow(airStations)
days <- index(NO2zoo)
nDays <- length(days)
## Duration in seconds of the animation
duration <- nDays*.3
```

The first snapshot of the data is produced with spplot. We define an auxiliary function, panel.circlesplot, to display the data encoding values with circles of variable size and color. This function uses grid.circle from the grid package.

The subsequent frames of the animation will modify the colors and sizes of the circles according to the NO2st object.

```
library(grid)

## Auxiliary panel function to display circles
panel.circlesplot <- function(x, y, cex, col = 'gray',
                    name = 'stationsCircles', ...){
    grid.circle(x, y, r = cex,
              gp = gpar(fill = col, alpha = 0.5),
              default.units = 'native', name = name)
}

pStart <- spplot(start, panel = panel.circlesplot,
              cex = startVals,
```

```
                     scales = list(draw = TRUE), auto.key = FALSE)
        pStart
```

17.2.2   Intermediate States to Create the Animation

From this initial state, grid.animate creates a collection of animated graphical objects with the intermediate states defined by animUnit and anim-Value. As previously stated, the NO_2 values will be encoded with the radius of each circle, and the color of the circles will distinguish between weekdays and weekend. The use of rep=TRUE ensures that the animation will be repeated indefinitely.

```
## Color to distinguish between weekdays ('green') and weekend
## ('blue')
isWeekend <- function(x) {format(x, '%w') %in% c(0, 6)}
color <- ifelse(isWeekend(days), 'blue', 'green')
colorAnim <- animValue(rep(color, each = nStations),
                id = rep(seq_len(nStations), nDays))

## Intermediate sizes of the circles
vals <- NO2st$vals/5000
vals[is.na(vals)] <- 0
radius <- animUnit(unit(vals, 'native'),
                id = rep(seq_len(nStations), nDays))

## Animation of circles including sizes and colors
grid.animate('stationsCircles',
        duration = duration,
        r = radius,
        fill = colorAnim,
        rep = TRUE)
```

17.2.3   Time Reference: Progress Bar

Information from an animation is better understood if a time reference is included, for example with a progress bar. The following code builds a progress bar with ticks at the first day of each month, and with color changing from gray (background) to blue as the time advances. On the other hand, it is convenient to provide a method so the user can stop and restart the animation sequence if desired. This functionality is added with the definition of two events, onmouseover and onmouseout, included with the grid.garnish function.

```
## Progress bar
prettyDays <- pretty(days, 12)
## Width of the progress bar
pbWidth <- .95
## Background
grid.rect(.5, 0.01, width = pbWidth, height = .01,
        just = c('center', 'bottom'),
        name = 'bgbar', gp = gpar(fill = 'gray'))

## Width of the progress bar for each day
dayWidth <- pbWidth/nDays
ticks <- c(0, cumsum(as.numeric(diff(prettyDays)))*dayWidth) +
    .025
grid.segments(ticks, .01, ticks, .02)
grid.text(format(prettyDays, '%d-%b'),
        ticks, .03, gp = gpar(cex = .5))
## Initial display of the progress bar
grid.rect(.025, .01, width = 0,
        height = .01, just = c('left', 'bottom'),
        name = 'pbar', gp = gpar(fill = 'blue', alpha = '.3'))
## ...and its animation
grid.animate('pbar', duration = duration,
            width = seq(0, pbWidth, length = duration),
            rep = TRUE)
## Pause animations when mouse is over the progress bar
grid.garnish('bgbar',
            onmouseover = 'document.rootElement.pauseAnimations()',
            onmouseout = 'document.rootElement.unpauseAnimations()'
                )
```

The SVG file is finally produced with grid.export (Figure 17.3)

```
grid.export('figs/NO2pb.svg')
```

17.2.4   Time Reference: A Time Series Plot

A different and more informative solution is to add a time series plot instead of a progress bar. This time series plot displays the average value of the set of stations, with a point and a vertical line to highlight the time position as the animation advances (Figure 17.4).

```
## Time series with average value of the set of stations
NO2mean <- zoo(rowMeans(NO2zoo, na.rm = TRUE), index(NO2zoo))
## Time series plot with position highlighted
```

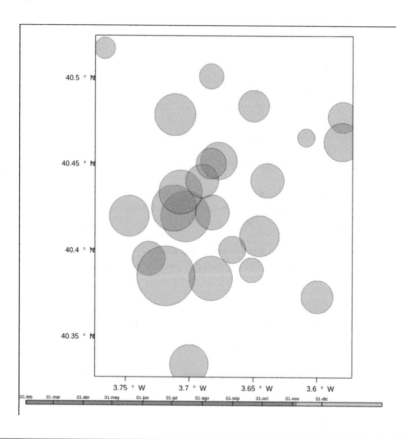

FIGURE 17.3: Animated circles of the NO_2 space-time data with a progress bar.

```
pTimeSeries <- xyplot(NO2mean, xlab = '', identifier = 'timePlot'
   ) +
  layer({
     grid.points(0, .5, size = unit(.5, 'char'),
                 default.units = 'npc',
                 gp = gpar(fill = 'gray'),
                 name = 'locator')
     grid.segments(0, 0, 0, 1, name = 'vLine')
  })
```

```
print(pStart, position = c(0, .2, 1, 1), more = TRUE)
print(pTimeSeries, position = c(.1, 0, .9, .25))
```

Once again, `grid.animate` creates a sequence of intermediate states for each object of the graphical scenes: The signaling point and vertical line follow the time evolution, while the sizes and colors of each station circle change as in the previous approach. Moreover, the onmouseover and onmouseout events are defined with `grid.garnish` so the user can pause and restart the animation by hovering the mouse over the time series plot.

```
grid.animate('locator',
             x = unit(as.numeric(index(NO2zoo)), 'native'),
             y = unit(as.numeric(NO2mean), 'native'),
             duration = duration, rep = TRUE)
```

```
xLine <- unit(index(NO2zoo), 'native')
```

```
grid.animate('vLine',
             x0 = xLine, x1 = xLine,
             duration = duration, rep = TRUE)
```

```
grid.animate('stationsCircles',
             duration = duration,
             r = radius,
             fill = colorAnim,
             rep = TRUE)
```

```
## Pause animations when mouse is over the time series plot
grid.garnish('timePlot', grep = TRUE,
             onmouseover = 'document.rootElement.pauseAnimations()',
             onmouseout = 'document.rootElement.unpauseAnimations()'
             )
```

```
grid.export('figs/vLine.svg')
```

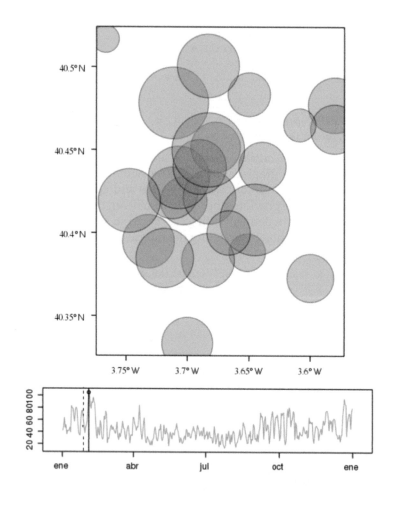

FIGURE 17.4: Animated circles of the NO_2 space-time data with a time series as reference.

17.3 Fly-by Animation

In this section the rgl package is used to generate a fly-by animation over the Earth, as an example of an animation depicting changes of a spatial attribute.

Basic 3D Earth

Firstly, a basic 3D Earth at night is created with the surface3d function, using the night lights images published by the NASA[6].

```
library(rgl)
library(magick) ## needed to import the texture

## Opens the OpenGL device with a black background
open3d()
bg3d('black')

## XYZ coordinates of a sphere
lat <- seq(-90, 90, len = 100) * pi/180
long <- seq(-180, 180, len = 100) * pi/180
r <- 6378.1 # radius of Earth in km
x <- outer(long, lat, FUN = function(x, y) r * cos(y) * cos(x))
y <- outer(long, lat, FUN = function(x, y) r * cos(y) * sin(x))
z <- outer(long, lat, FUN = function(x, y) r * sin(y))

## Read, scale, and convert the image
nightLightsJPG <- image_read("https://eoimages.gsfc.nasa.gov/
    images/imagerecords/79000/79765/dnb_land_ocean_
    ice.2012.13500x6750.jpg")
nightLightsJPG <- image_scale(nightLightsJPG, "8192") ##
    surface3d reads files up to 8192x8192
nightLights <- image_write(nightLightsJPG, tempfile(),
                    format = 'png') ## Only the png format is
                                    supported
## Display the sphere with the image superimposed
surface3d(-x, -z, y,
        texture = nightLights,
        specular = "black", col = 'white')
```

---

[6]The page "Out of the Blue and Into the Black: New Views of the Earth at Night", https://earthobservatory.nasa.gov/Features/IntotheBlack/, provides detailed information about the Earth at Night maps.

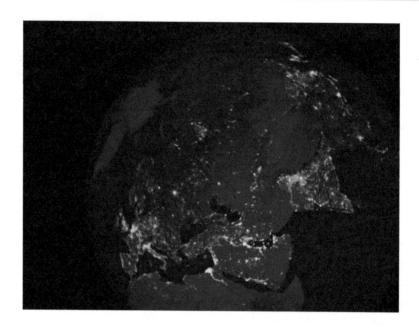

FIGURE 17.5: Snapshot of the WebGL figure created with `writeWebGL`.

This OpenGL object can be exported to different formats. For example, Figure 17.5 shows a snapshot of the WebGL figure created with `writeWebGL`:

```
writeWebGL('nightLights', width = 1000)
```

Define the Locations

Once the Earth is represented with the sphere and the superimposed image, the fly-by animation is defined with a set of locations to be visited:

```
cities <- rbind(c('Madrid', 'Spain'),
                c('Tokyo', 'Japan'),
                c('Sidney', 'Australia'),
                c('Sao Paulo', 'Brazil'),
                c('New York', 'USA'))
cities <- as.data.frame(cities)
names(cities) <- c("city", "country")
```

The latitude and longitude coordinates of these cities can be obtained through the Nominatim service of OpenStreetMap. An auxiliary function, geocode, obtains this information using the XML package.

```
library(XML)

geocode <- function(x){
    city <- x[1]
    country <- x[2]
    urlOSM <- paste0('http://nominatim.openstreetmap.org/search?',
                     'city=', city,
                     '&country=', country,
                     '&format=xml')
    ## Parse the webpage
    xmlOSM <- xmlParse(urlOSM)
    ## Use only the first result
    cityOSM <- getNodeSet(xmlOSM, '//place')[[1]]
    ## Extract attributes: longitude...
    lon <- xmlGetAttr(cityOSM, 'lon')
    ## and latitude
    lat <- xmlGetAttr(cityOSM, 'lat')
    ## Return them as a vector
    as.numeric(c(lon, lat))
}

points <- apply(cities, 1, geocode)
points <- t(points)
colnames(points) <- c("lon", "lat")

cities <- cbind(cities, points)
```

Generate the Route

The next step computes the intermediate points between each pair of locations. The geosphere package provides the gcIntermediate function for this task:

```
library(geosphere)

## When arriving or departing include a progressive zoom with 100
## frames
zoomIn <- seq(.3, .1, length = 100)
zoomOut <- seq(.1, .3, length = 100)
```

```
## First point of the route
route <- data.frame(lon = cities[1, "lon"],
                    lat = points[1, "lat"],
                    zoom = zoomIn,
                    name = cities[1, "city"],
                    action = 'arrive')

## This loop visits each location included in the 'points' set
## generating the route.
for (i in 1:(nrow(cities) - 1)) {

    p1 <- cities[i,]
    p2 <- cities[i + 1,]
    ## Initial location
    departure <- data.frame(lon = p1$lon,
                            lat = p1$lat,
                            zoom = zoomOut,
                            name = p1$city,
                            action = 'depart')

    ## Travel between two points: Compute 100 points between the
    ## initial and the final locations.
    routePart <- gcIntermediate(p1[, c("lon", "lat")],
                     p2[, c("lon", "lat")],
                     n = 100)
    routePart <- data.frame(routePart)
    routePart$zoom <- 0.3
    routePart$name <- ''
    routePart$action <- 'travel'

    ## Final location
    arrival <- data.frame(lon = p2$lon,
                          lat = p2$lat,
                          zoom = zoomIn,
                          name = p2$city,
                          action = 'arrive')
    ## Complete route: initial, intermediate, and final locations.
    routePart <- rbind(departure, routePart, arrival)
    route <- rbind(route, routePart)
}

## Close the travel
route <- rbind(route,
```

```
            data.frame(lon = cities[i + 1, "lon"],
                       lat = cities[i + 1, "lat"],
                       zoom = zoomOut,
                       name = cities[i+1, "city"],
                       action = 'depart'))
```

`summary(route)`

```
      lon lat zoom name
Min.  :-179.538 Min.  :-74.346 Min.  :0.1000 Madrid  :300
1st Qu.: -54.003 1st Qu.:-23.551 1st Qu.:0.1707 New York  :400
Median : -3.704 Median : 25.285 Median :0.2434 Sao Paulo:400
Mean  : 32.888 Mean  : 6.293 Mean  :0.2296 Sidney  :400
3rd Qu.: 139.759 3rd Qu.: 35.683 3rd Qu.:0.3000 Tokyo  :400
Max.  : 178.515 Max.  : 68.234 Max.  :0.3000  :800
   action
arrive: 900
depart:1000
travel: 800
```

Produce the Frames

Finally, this matrix of points is used to change the viewpoint of the OpenGL scene with the rgl.viewpoint function. The travel function wraps this function to automate the process with the movie3d function. Figure 17.6 shows an example of a frame produced with this function.

```
## Function to move the viewpoint in the RGL scene according to
   the
## information included in the route (position and zoom).
travel <- function(tt){
  point <- route[tt,]
  rgl.viewpoint(theta = -90 + point$lon,
            phi = point$lat,
            zoom = point$zoom)
}

## Example of usage of travel
## Frame no.1200
travel(1200)
rgl.snapshot(figs/rgl_travel1200.pdf')
```

The movie3d accepts a function, travel in our code, to modify the RGL scene. It creates an snapshot at each step, and paste these snapshots as frames of a movie.

FIGURE 17.6: Example of usage of the `travel` function (frame no.1200).

```
movie3d(travel,
        duration = nrow(route),
        startTime = 1, fps = 1,
        type = 'mp4', clean = FALSE)
```

Part IV

Glossary, Bibliography and Index

Glossary

CM-SAF: Satellite Application Facility on Climate Monitoring.

CRAN: Comprehensive R Archive Network.

DEM: Digital Elevation Model.

Geo-TIFF: A public domain metadata standard which allows georeferencing information to be embedded within a TIFF file.

GeoJSON: Format for encoding a variety of geographic data structures.

GIS: Geographic Information Systems.

IDW: Inverse Distance Weighted interpolation.

KML: Keyhole Markup Language, an XML notation for expressing geographic annotation and visualization within Internet-based, two-dimensional maps and three-dimensional Earth browsers.

NEO-NASA: NASA Earth Observations, part of the NASA's Earth Observing System (EOS).

NetCDF: Network Common Data Form, a set of software libraries and self-describing, machine-independent data formats that support the creation, access, and sharing of array-oriented scientific data.

RasterBrick: A class to represent multilayer (variable) raster data.

RasterLayer: A class to represent single-layer (variable) raster data.

RasterStack: A class to represent multilayer (variable) raster data.

shapefile: A spatial data format. It stores geometry and attribute information for the spatial features in a data set.

SIS: Shortwave incoming solar radiation.

SpatialLinesDataFrame: Class for spatial attributes consisting of sets of lines, where each set of lines relates to an attribute row in a data.frame.

SpatialPixelsDataFrame: Class for spatial attributes that have spatial locations on a regular grid.

SpatialPointsDataFrame: Class for spatial attributes that have spatial point locations.

SpatialPolygonsDataFrame: Class to hold polygons with attributes.

STL: File format that encodes the surface geometry of a 3D object using tessellation.

SVG: Scalable Vector Graphics.

TIFF: Tagged Image File Format, a computer file format for storing raster graphics images.

WebGL: Web Graphics Library, a JavaScript API for rendering interactive 2D and 3D graphics within any compatible web browser without the use of plugins.

WRF: Weather Research and Forecasting model.

XML: Extensible Markup Language, a markup language that defines a set of rules for encoding documents in a format that is both human-readable and machine-readable.

Bibliography

Antonanzas-Torres, F., F. Cañizares, and O. Perpiñán (2013). "Comparative assessment of global irradiation from a satellite estimate model (CM SAF) and on-ground measurements (SIAR): A Spanish case study". In: *Renewable and Sustainable Energy Reviews* 21, pp. 248–261. ISSN: 1364-0321. DOI: 10.1016/j.rser.2012.12.033. URL: https://github.com/oscarperpinan/CMSAF-SIAR.

Becker, R. A., A. R. Wilks, and R. Brownrigg (2017). *mapdata: Extra Map Databases*. R package version 2.2-6. URL: http://CRAN.R-project.org/package=mapdata.

Becker, R. A., A. R. Wilks, R. Brownrigg, et al. (2017). *maps: Draw Geographical Maps*. R package version 3.2.0. URL: http://CRAN.R-project.org/package=maps.

Bivand, R. (2017). *classInt: Choose Univariate Class Intervals*. R package version 0.1-24. URL: http://CRAN.R-project.org/package=classInt.

Bivand, R., T. Keitt, and B. Rowlingson (2017). *rgdal: Bindings for the Geospatial Data Abstraction Library*. R package version 1.2-16. URL: http://CRAN.R-project.org/package=rgdal.

Bivand, R. and N. Lewin-Koh (2017). *maptools: Tools for Reading and Handling Spatial Objects*. R package version 0.9-2. URL: http://CRAN.R-project.org/package=maptools.

Bivand, R., E. J. Pebesma, and V. Gomez-Rubio (2013). *Applied Spatial Data Analysis with R*. Springer, New York. URL: http://www.asdar-book.org/.

Byron, L. and M. Wattenberg (2008). *Stacked Graphs – Geometry & Aesthetics*. Tech. rep. URL: http://www.leebyron.com/else/streamgraph/download.php?file=stackedgraphs_byron_wattenberg.pdf.

Cairo, A. (2012). *The Functional Art: An Introduction to Information Graphics and Visualization*. New Riders Publishing, Aarhus, Denmark.

Carr, D. B. et al. (1987). "Scatterplot Matrix Techniques for Large N". English. In: *Journal of the American Statistical Association* 82.398, pp. 424–436. ISSN: 01621459. URL: http://www.jstor.org/stable/2289444.

Carr, D., N. Lewin-Koh, and M. Maechler (2018). *hexbin: Hexagonal Binning Routines*. R package version 1.27.2. URL: https://CRAN.R-project.org/package=hexbin.

Chambers, J. M. (2008). *Software for Data Analysis: Programming with R*. ISBN 978-0-387-75935-7. Springer, New York.
URL: http://stat.stanford.edu/~jmc4/Rbook/.

Chatfield, C. (2016). *The Analysis of Time Series: An Introduction*. Chapman & Hall/CRC Texts in Statistical Science. CRC Press. ISBN: 9780203491683.

Cleveland, W. S. (1993). *Visualizing Data*. Summit, NJ: Hobart Press.

— (1994). *The Elements of Graphing Data*. Murray Hill, NJ.: AT&T, Bell Laboratories.

Cleveland, W. S. and R. McGill (1984). "Graphical Perception: Theory, Experimentation, and Application to the Development of Graphical Methods". In: *Journal of the American Statistical Association* 79.387, pages. ISSN: 01621459. URL: http://www.jstor.org/stable/2288400.

CM SAF (2013). *The Satellite Application Facility on Climate Monitoring*. http://www.cmsaf.eu.

Cressie, N. and C.K. Wikle (2015). *Statistics for Spatio-Temporal Data*. Wiley Series in Probability and Statistics. Wiley, New York. ISBN: 9781119243045.

Dent, B., J. Torguson, and T. Hodler (2008). *Cartography: Thematic Map Design*. McGraw-Hill Education, New York. ISBN: 9780072943825.

Few, S. (2007). *Visualizing Change: An Innovation in Time-Series Analysis*. Tech. rep. Perceptual Edge, Berkeley, CA.
URL: http://www.perceptualedge.com/articles/visual_business_intelligence/visualizing_change.pdf.

— (2008). *Time on the Horizon*. Tech. rep. Perceptual Edge, Berkeley, CA.
URL: http://www.perceptualedge.com/articles/visual_business_intelligence/time_on_the_horizon.pdf.

Friendly, M. and D. Denis (2005). "The early origins and development of the scatterplot". In: *Journal of the History of the Behavioral Sciences* 41.2, pp. 103–130. ISSN: 1520-6696. DOI: 10.1002/jhbs.20078.

Gesmann, M. and D. de Castillo (2011). "googleVis: Interface between R and the Google Visualisation API". In: *The R Journal* 3.2, pp. 40–44. URL:

http://journal.r-project.org/archive/2011-2/RJournal_2011-2_Gesmann+de~Castillo.pdf.

Grothendieck, G. and T. Petzoldt (2004). "R Help Desk: Date and Time Classes in R". In: *R News* 4.1, pp. 29–32. URL: http://CRAN.R-project.org/doc/Rnews/Rnews_2004-1.pdf.

Harrower, M. and S. I. Fabrikant (2008). "The Role of Map Animation in Geographic Visualization". In: *Geographic Visualization: Concepts, Tools and Applications*. Ed. by M. Dodge, M. McDerby, and Turner M. Chichester, UK: Wiley, pp. 49–65. URL: http://www.zora.uzh.ch/8979/.

Havre, S. et al. (2002). "ThemeRiver: Visualizing Thematic Changes in Large Document Collections". In: *IEEE Transactions on Visualization and Computer Graphics* 8.1, pp. 9–20. ISSN: 1077-2626. DOI: 10.1109/2945.981848.

Heer, J. and M. Agrawala (2006). "Multi-Scale Banking to 45 Degrees". In: *IEEE Transactions on Visualization and Computer Graphics* 12.5, pp. 701–708. ISSN: 1077-2626. DOI: 10.1109/TVCG.2006.163. URL: http://vis.berkeley.edu/papers/banking/2006-Banking-InfoVis.pdf.

Heer, J., M. Bostock, and V. Ogievetsky (2010). "A tour through the visualization zoo". In: *Communications of the ACM* 53.6, pp. 59–67. ISSN: 0001-0782. DOI: 10.1145/1743546.1743567. URL: http://doi.acm.org/10.1145/1743546.1743567.

Heer, J., N. Kong, and M. Agrawala (2009). "Sizing the Horizon: The Effects of Chart Size and Layering on the Graphical Perception of Time Series Visualizations". In: *ACM Human Factors in Computing Systems (CHI)*, pp. 1303–1312. URL: http://vis.berkeley.edu/papers/horizon/2009-TimeSeries-CHI.pdf.

Hengl, T. (2009). *A Practical Guide to Geostatistical Mapping*. University of Amsterdam, Amsterdam. URL: http://spatial-analyst.net/book/.

Hijmans, R. J. (2017). *raster: Geographic Data Analysis and Modeling*. R package version 2.6-7. URL: https://CRAN.R-project.org/package=raster.

Hocking, T. D. (2017). *directlabels: Direct Labels for Multicolor Plots*. R package version 2017.03.31. URL: http://CRAN.R-project.org/package=directlabels.

Hovmöller, E. (1949). "The Trough-and-Ridge diagram". In: *Tellus* 1.2, pp. 62–66. ISSN: 2153-3490. DOI: 10.1111/j.2153-3490.1949.tb01260.x.

Ihaka, R. et al. (2016). *colorspace: Color Space Manipulation*. R package version 1.3-2. URL: http://CRAN.R-project.org/package=colorspace.

Kahle, D. and H. Wickham (2013). "ggmap: Spatial Visualization with gg-plot2". In: *The R Journal* 5.1, pp. 144–161. URL: http://journal.r-project.org/archive/2013-1/kahle-wickham.pdf.

Kropotkin, P. (1906). *The Conquest of Bread.* G. P. Putnam's Sons. URL: http://en.wikisource.org/wiki/The_Conquest_of_Bread.

Lucchesi, L. R. and C. K. Wikle (2017). "Visualizing uncertainty in areal data with bivariate choropleth maps, map pixelation and glyph rotation". In: *Stat* 6.1, pp. 292–302. DOI: 10.1002/sta4.150.

MARM (2011). *Sistema de Información Agroclimática del Regadío.* http://www.marm.es/siar/Informacion.asp.

McIlroy, D. et al. (2017). *mapproj: Map Projections.* R package version 1.2-5. URL: http://CRAN.R-project.org/package=mapproj.

Meihoefer, H. J. (1969). "The Utility of the Circle as an Effective Carto-graphic Symbol". In: *Cartographica: The International Journal for Geographic Information and Geovisualization* 6 (2), pp. 104–117. DOI: 10.3138/J04Q-1K34-26X1-7244.

Mumford, L. (1934). *Technics and Civilization.* San Diego, CA.: Harcourt, Brace & Company, Inc.

Murrell, P. (2011). *R Graphics.* 2nd. The R Series. Boca Raton, FL.: Chapman & Hall/CRC, p. 546.

Murrell, P. and S. Potter (2017). *gridSVG: Export grid graphics as SVG.* R package version 1.6-0. URL: http://CRAN.R-project.org/package=gridSVG.

Neuwirth, E. (2014). *RColorBrewer: ColorBrewer Palettes.* R package version 1.1-2. URL: http://CRAN.R-project.org/package=RColorBrewer.

Pebesma, E. (2012). "spacetime: Spatio-Temporal Data in R". In: *Journal of Statistical Software* 51.7, pp. 1–30. ISSN: 1548-7660. URL: http://www.jstatsoft.org/v51/i07.

— (2018). *sf: Simple Features for R.* R package version 0.6-0. URL: https://CRAN.R-project.org/package=sf.

Pebesma, E. J. (2004). "Multivariable Geostatistics in S: The gstat Package". In: *Computers and Geosciences* 30, pp. 683–691.

Pebesma, E. J. and R. Bivand (2005). "Classes and methods for spatial data in R". In: *R News* 5.2, pp. 9–13. URL: http://CRAN.R-project.org/doc/Rnews/.

Perpiñán, O. (2012). "solaR: Solar Radiation and Photovoltaic Systems with R". In: *Journal of Statistical Software* 50.9, pp. 1–32. URL: http://www.jstatsoft.org/v50/i09/.

Perpiñán, O. and M. P. Almeida (2015). "meteoForecast: Numerical Weather Predictions". In: DOI: 10.5281/zenodo.13882. URL: https://github.com/oscarperpinan/meteoForecast.

Perpiñán, O. and R. Hijmans (2017). *rasterVis: Visualization Methods for the raster Package*. R package version 0.43. URL: http://oscarperpinan.github.io/rastervis/.

Pinho-Almeida, M., O. Perpiñán, and L Narvarte (2015). "PV Power Forecast Using a Nonparametric PV Model". In: *Solar Energy* 115, pp. 354–368. ISSN: 0038092X. DOI: 10.1016/j.solener.2015.03.006. URL: oscarperpinan.github.io/papers/Pinho.Perpinan.ea2014.pdf.

Posselt, R., R.W. Mueller, et al. (2012). "Remote sensing of solar surface radiation for climate monitoring — the CM SAF retrieval in international comparison". In: *Remote Sensing of Environment* 118, pp. 186–198. ISSN: 0034-4257. DOI: 10.1016/j.rse.2011.11.016.

Posselt, R., R. Müller, et al. (2011). *CM SAF Surface Radiation MVIRI Data Set 1.0 - Monthly Means / Daily Means / Hourly Means*. DOI: 10.5676/EUM_SAF_CM/RAD_MVIRI/V001. URL: http://dx.doi.org/10.5676/EUM_SAF_CM/RAD_MVIRI/V001.

R Development Core Team (2017). *R: A Language and Environment for Statistical Computing*. R Foundation for Statistical Computing. Vienna, Austria. URL: http://www.R-project.org.

Ripley, B. D. and K. Hornik (2001). "Date-Time Classes". In: *R News* 1.2, pp. 8–11. URL: http://CRAN.R-project.org/doc/Rnews/Rnews_2001-2.pdf.

Rossini, A. J. et al. (2004). "Emacs Speaks Statistics: A Multiplatform, Multipackage Development Environment for Statistical Analysis". In: *Journal of Computational and Graphical Statistics* 13.1, pp. 247–261. DOI: 10.1198/1061860042985. eprint: http://www.tandfonline.com/doi/pdf/10.1198/1061860042985. URL: http://ess.r-project.org/.

Ryan, J. A. and J. M. Ulrich (2013). *xts: eXtensible Time Series*. R package version 0.9-5. URL: http://CRAN.R-project.org/package=xts.

Sarkar, D. (2008). *Lattice: Multivariate Data Visualization with R*. ISBN 978-0-387-75968-5. New York: Springer. URL: http://lmdvr.r-forge.r-project.org.

Sarkar, D. and F. Andrews (2016). *latticeExtra: Extra Graphical Utilities Based on lattice*. R package version 0.6-28. URL: http://CRAN.R-project.org/package=latticeExtra.

Schulte, E. et al. (2012). "A Multi-Language Computing Environment for Literate Programming and Reproducible Research". In: *Journal of Sta-*

tistical Software 46.3, pp. 1–24. ISSN: 1548-7660. URL: http://www.jstatsoft.org/v46/i03.

Schulz, J. et al. (2009). "Operational Climate Monitoring from Space: The EUMETSAT Satellite Application Facility on Climate Monitoring". In: *Atmospheric Chemistry and Physics* 9, pp. 1687–1709. DOI: 10.5194/acp-9-1687-2009. URL: http://www.atmos-chem-phys.net/9/1687/2009/acp-9-1687-2009.pdf.

Slocum, T. A. (2005). *Thematic Cartography and Geographic Visualization*. Englewood Cliffs, NJ.: Prentice Hall. ISBN: 9780130351234.

Tufte, E. R. (1990). *Envisioning information*. Cheshire, CT.: Graphic Press.

— (2001). *The Visual Display of Quantitative Information*. Cheshire, CT.: Graphic Press.

Vaidyanathan, R. et al. (2017). *htmlwidgets: HTML Widgets for R*. R package version 0.9. URL: https://CRAN.R-project.org/package=htmlwidgets.

Ware, C. (2008). *Visual Thinking for Design*. Burlington, MA.: Morgan Kaufmann Pub. ISBN: 9780123708960.

Wegenkittl, R. and E. Gröller (1997). "Fast Oriented Line Integral Convolution for Vector Field Visualization via the Internet". In: *Proceedings of the 8th Conference on Visualization'97*. IEEE Computer Society Press, pp. 309–316. URL: http://christl.cg.tuwien.ac.at/research/vis/dynsys/frolic/frolic_crc.pdf.

Wickham, H. (2016). *ggplot2: Elegant Graphics for Data Analysis*. Springer. URL: https://github.com/hadley/ggplot2-book/.

Wilkinson, L. (2005). *The Grammar of Graphics*. Springer.

Wills, G. (2011). *Visualizing Time: Designing Graphical Representations for Statistical Data*. Statistics and Computing. New York: Springer. ISBN: 9780387779065.

Xie, Y. (2013). "animation: An R Package for Creating Animations and Demonstrating Statistical Methods". In: *Journal of Statistical Software* 53.1, pp. 1–27. ISSN: 1548-7660. URL: http://www.jstatsoft.org/v53/i01.

Zeileis, A. and G. Grothendieck (2005). "zoo: S3 Infrastructure for Regular and Irregular Time Series". In: *Journal of Statistical Software* 14.6, pp. 1–27. URL: http://www.jstatsoft.org/v14/i06/.

Index